CIRCULAR AGRICULTURE

AS A CLIMATE-SMART TOOL FOR CIRCULAR ECONOMIES

DR. ISAAC KWADWO MPANGA

Circular Planet Institute LLC (Founder)

Table of Contents

Executive Summary

Feeding and sheltering the increasing world population and ensuring food security for all is complex, especially when our current linear agriculture, manufacturing practices, and other related human activities are rapidly depleting the limited natural resources and increasing global warming at an unsustainable rate. The urgency to transition from the current dominant linear agricultural systems to a more circular agriculture (CA) with strategic recycling and regenerative practices has never been stronger nor timelier than today—giving us the opportunity to act passionately using the available science-based research information and technology. However, we must acknowledge that it is a sensitive and complicated process due to differences in agricultural practices, manufacturing, local community needs, governments' priorities, limited research, and technology. Scaling up and accelerating CA with these complexities requires implementing it at distinct levels that accounts for these complexities that would lead to ownership by the varied stakeholders. Current models are limited in terms of these distinct levels at which CA can be implemented while accounting for the complex stakeholders and factors. This book discusses a model that would help implement CA at distinct levels such as single farm, community, region/state, country, and global, while accounting for the variations in stakeholders, resources, relationships, productivity, and efficiencies in meeting climate-smart targets at each level. Also, the principles, practices, and types of CA are elaborated with emphasis on the critical roles composting, cover crops, crop rotations, biofertilizers, integrated pest management, solar farming (agrivoltaic-food and solar energy production on the same land at the same time), etc. This book also discusses and recognizes the role of research, technology, governance, industry and manufacturing, farmers, society/culture, and consumers as major stakeholders and drive scaling CA. Identifying all partners as equal players with ownership, especially the end-product users and local communities are discussed as a crucial catalyst for transition through the promotion of science-based approaches and innovative technologies. Finally, realization that CA as a climate-smart tool will meet a level of resistance is emphasized in this book as an important understanding to have in local context to help respect all partners as equal players with one goal for a bigger impact towards circularity of local and global economies.

Acknowledgement

Glory be to God Almighty for His grace, provisions, strength, knowledge, and good health. I thank my family, especially my wife and son (Sabrina and Josiah Taasun), mother and father (Grace and Thomas Bignabil), brother (Pastor Daniel), and sisters (Veronica, Agnes, Abena, and Mansa) sincerely for their dedicated prayers, support, and inspiration that keeps me going every day. My sincere thank you to Professor Harrison Dapaah and Professor Guenter Neumann in Ghana and Germany, respectively— both supported my professional journey, and I would not be where I am today without their guidance and support. Also, to all partners, co-authors, institutions/companies (PepsiCo, North America; University of Arizona, USA; University of Hohenheim, Germany, University of Energy and Natural Resources, Ghana; and Ghana Education Services, Ghana), and communities I worked with in the past or am working with at the time of writing this book, thank you for the challenge and opportunity to have served you. Finally, thank you to the Church of Pentecost (pastors, elders, and members) for their prayers as well as all friends and families who encourages and motivates me every day.

Preface

This book is a comprehensive guide for farmers, companies not-for-profit companies, consultants, extension professionals, and the public with interest in climate solution to reduce greenhouse gas emissions, increase biodiversity, improve our health and wellness, and improve our livelihood with food security and climate resilience. This book shares my experiences and passion in sustainable agriculture, which I gained growing up on a farm in Ghana and from my professional journey that has limited space in the traditional scientific journals, where most of my scientific work is published. Growing up on a farm with my parents, I was the one who fed the chickens using waste from local mills and a by-product called "pito chaff", collected from home-made breweries. I also cleaned the chicken coop of droppings and gathered it as manure for my mom to use as a fertilizer on her vegetables. I saw my father use legume crops after maize and yam, mixed cropped legumes and cereal or tubers on the same field, mixed farm crops and livestock (I can still picture chickens all over the fields to pick up pests and worms), with no synthetic inputs. In my studies, research, consulting, and farming experiences, I know and believe in circular agriculture's potential as a climate solution, and all we need to do is be responsible, stop waiting for tomorrow or someone else, and act today at our level.

This book presents a comprehensive guide on how to implement circular agriculture using Mpanga's hierarchical implementation model, the principles and practices of circular agriculture, how to select and implement a circular agriculture that fits your needs or stakeholder's needs and their capacity at different levels, the role of stakeholders (farmers, manufacturers, researchers, government, and community) in scaling CA, and how to measure progress. The role of compost, bio-fertilizers, agrivoltaic, livestock, integrated pest management, cover crops, no-till, mixed farming, and cropping are presented with opportunities for improvement and how to scale with ownership.

CHAPTER 1

Introduction

The world's human population is projected at over nine billion by 2050, with the responsibility of producing 70 percent more food on the same fixed land asset and limited resources. The growth in human population and the accompanying anthropogenic activities is stretching our natural resources, such as water and other resources for agriculture (Fig. 1a and b) and have left a negative impact on the climate with increasing drought, wildfires, floods, temperatures, decreasing biodiversity and more, which limits agricultural production abilities [1].

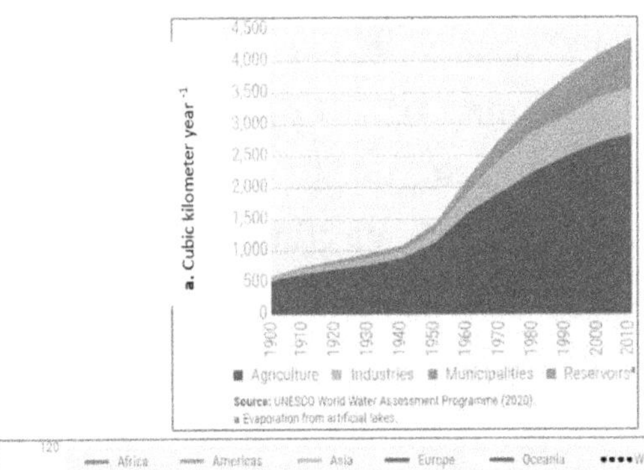

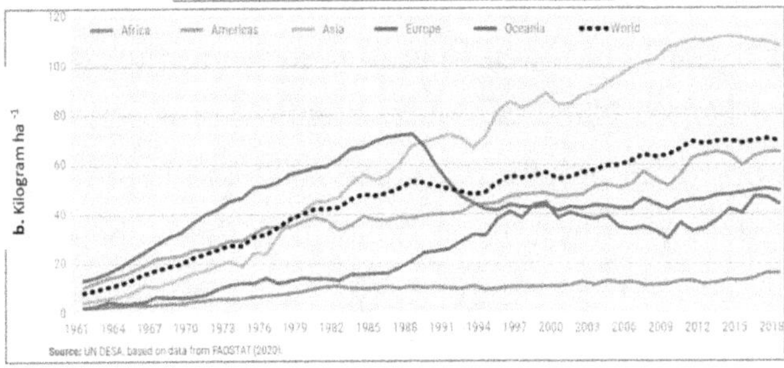

*Figure 1: Global water withdrawal from 1990 to 2010 (**a**) and nitrogen fertilizer use by region per cropland area from 1961 to 2018 (**b**) (Source: Used with permission from the United Nations, Department of Economic and Social Affairs Economic Analysis) [2].*

To meet the food demands of our growing population in the past and current, the focus has been on intensive and linear agriculture systems that heavily depend on natural resources, such as water (Fig. 1a), and external inputs, such as synthetic chemicals (Fig. 1b), for productivity, without much conservative effort, but, rather, increasing negative effects on the ecosystem. The devastating effects of current agricultural systems on our environment and health has a direct relationship to the linearity of agricultural practices that rely on heavy intensive tillage, the over use of synthetic inputs such as nitrogen (Fig. 1b). These linear systems have many waste leakages from the suppliers of inputs and on the farm with disastrous effects on soil losses. These leakages have to do with both inefficiencies of the synthetic inputs, wrong applications (amount, time, and methods), tillage practices, and inability to return the organic nutrients removed from the field. For example, Anas, et. al. (2020) reported a varied nitrogen use efficiency (product of nitrogen uptake efficiency and nitrogen utilization efficiency) between 30 to 53 percent, which means nitrogen losses ranged from 70 to 47 percent [3]. Most of this nitrogen losses in the agricultural fields due to inefficiencies ends up in our grand waters as potential pollutants, affecting biodiversity, and the general ecosystem. In a recent study, it is revealed that 64 percent of global agricultural arable lands are at risk of pesticides pollution with high risk to the environmental and human health [4], which is extremely disturbing that their use will even continue to increase in the coming years. According to a United Nations report, chemical production for agriculture doubled up to 2.3 billion tons since 2000 and is projected to again double in 2030. Yes, there is no doubt that these chemicals helped increased our agricultural productivity, but we must and can use them responsibly. The proposed 70 percent extra food production by 2050 to feed the expected over nine billion human population does not seem sustainable while food waste is on the rise. Reducing, reusing, and recycling waste is key to ending hunger, and we must do better in this regard "at good will" with the current advances in technology and innovations at hand.

According to a report published by EPA in 2020 on estimated wasted food, 103 million tons of wasted food were generated in U.S.A. in 2018 from industrial (39 percent), residential (24 percent), restaurants (17 percent), commercial (12 percent), and institutional sectors (5 percent) and other sectors (3 percent) (Table 1) [5].

Table 1: Food wasted in the U.S.A. in 2018 by sector (EPA 2018 Food Waste report) [5]

Sectors food wasted in 2018 (total=103 million tons)	Percentage
Industry	39
Residential	24
Restaurants	17
Commercial (wholesale and supermarket)	12
Institutions	5
Others	3

A good amount of wasted food in the U.S.A. ended in the landfills, especially non-industrial sectors (Fig. 2a & b). The EPA report acknowledges that the majority of industry waste goes to land applications and animal feeds, which is why in figure 2a that includes management from industry has less landfill (36 percent) compared to figure 2b without industrial sector having landfill (56 percent) as a management strategy [5].

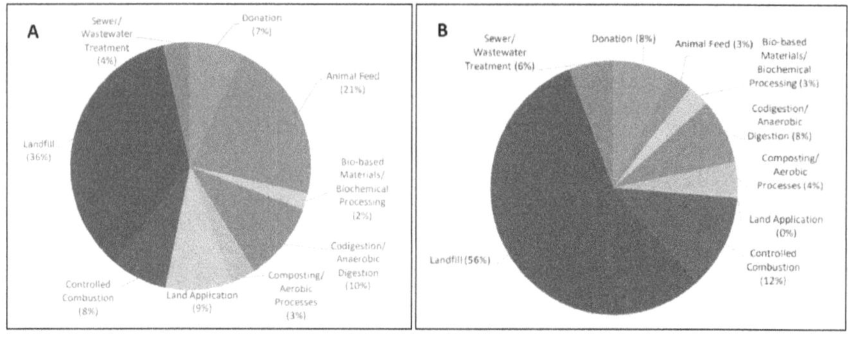

Figure 2: Percentage distribution of current food wasted management strategies by sector. A- including industry, B- excluding industry (EPA, 2018 Food Waste Report) [5].

Agricultural systems need to align and mimic nature's potential through biodiversity to reduce external inputs and natural resources leakages. Agricultural systems that are innovative and climate-smart such as circular agriculture with practices that are eco-friendly (improving soil health, carbon sequestration, and biodiversity while reducing greenhouse gas emissions) and has less to no waste streams needs more attention than ever before. According to Food and Agriculture Organization (FAO), biodiversity is a key resource in our effort to increase production systems and livelihoods resilient to shocks and stresses, including the effects of climate change and reduces our total dependence on synthetic inputs for agriculture productions [1]. Circular agriculture CA) is an agricultural system that recycles agricultural by-products, waste, and biomass from on- and off-farm facilities within a defined food system as renewable resources that helps limit the use of external inputs such as synthetic chemical fertilizers and pesticides (Fig. 3) [1,6].

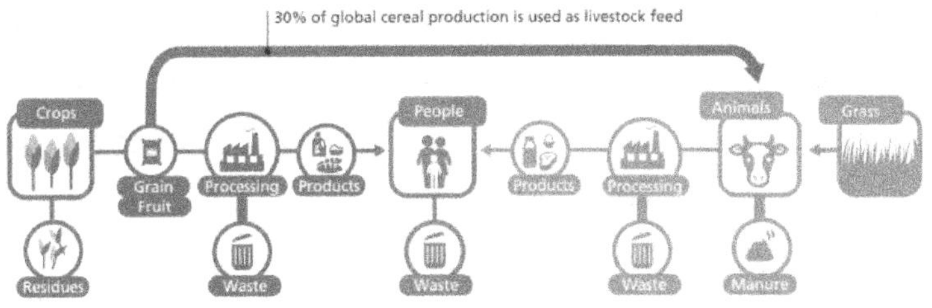

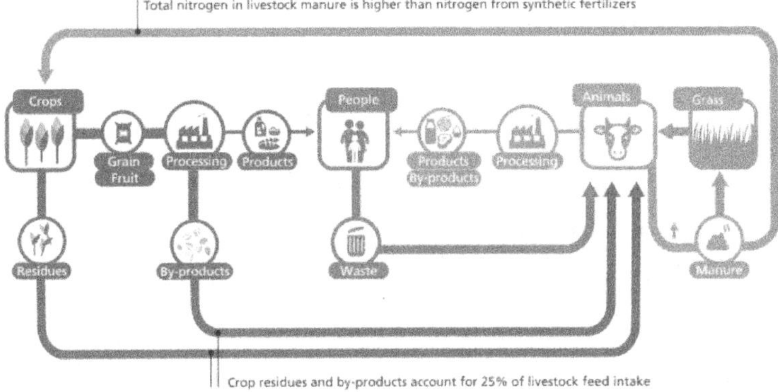

Source: FAO, 2018o.

Figure 3: *From linear to circular open-field agricultural system with livestock and grain integration (Sources: Food and Agriculture Organization of the United Nations. Reproduced with permission.) [1].*

Circular agriculture can be more "nature-inclusive agricultural systems." It is also termed as agroecological farming with specificity on ecosystem services such as nature and biodiversity use and retention on the farm landscape, which includes efficient and clean water use [6]. The aim of circular agriculture is to increase; 1) farm profitability, 2) environmentally friendly agricultural systems, and 3) animal and human welfare. Circular agriculture could enhance regenerative land-use systems with a reduction in chemical fertilizers and waste and could serve as a catalyst to global CO_2 emissions reduction and carbon sequestration. Circular agriculture is a broad system approach that could be practiced in many agricultural commodities, sectors, and geographical locations with strong contributions (directly or indirect) to most of the United Nations' sustainable development goals (SDGs). Most especially, CA relates to SDGs on no poverty (Goal 1), zero hunger (Goal 2), clean water and sanitation (Goal 6), responsible production and consumption (Goal 12), life on land (Goal 15). However, the specific practices would vary among commodities, sectors, and geographical locations, depending on the goals, practices, prevailing resources, culture, governance, and potential by-product streams. For example, the circular agricultural system in Verde Valley, Arizona, where vineyards and wineries are popular, will vary from circular agricultural practices in Yuma, Arizona, where leafy greens are mostly produced by growers and sent out of the state. Also, CA in my village [Borae No2. in Ghana, West Africa (in the context of developing country)] would be astronomically different from CA in Stuttgart, Germany (in the context of a developed country), due to many factors such as resources, climate, political and systems structures, financing, culture, etc. These complexities make CA implementation challenging such that one-fit-all approach would not work but presents an opportunity to look at the local context applications to scale it. In the next chapters, this book explores the concepts, principles, and levels of circular agriculture implementation and practical applications. I will also open the discussions on the collaborative role of farmers, manufacturing companies, financial institutions, communities, research, and governance. Other chapters discussed includes the role of compost, biofertilizers, integrated pest management (IPM), agrivoltaic (solar farming with agriculture production), and livestock in circular agriculture as climate-smart tools. Finally, scalability is discussed in the context of implementing CA at a hierarchical level and the key role of stakeholder involvement as players.

CHAPTER 2

Principles and Practices of Circular Agriculture

Circular agriculture is nothing new, but reemerging or renaming prominent ancient agricultural practices that were popular among pre-industrial societies that mimicked the natural ecosystem nutrient cycling through biodiversity and waste recycling with less negative impact to the environment (Fig. 3). Circular agriculture emphasizes promotion of smallholder and medium-scale farming such as organics, mixed cropping, crop rotation, waste reuse, and agroforestry practices [1,7]. In practice, a wider scale acceptance and use of these regenerative and circular agricultural practices is common among smallholder and medium farms depending less on synthetic inputs [7]. For example, in Europe, estimates show that a circular approach to food systems could reduce dependence on chemical fertilizers by 80 percent [2].

Does it mean we are returning to the old ways of farming with more manual work, less technology, and fewer innovations? The response is NO. According to Wageningen University in the Netherlands, Circular Agriculture is not; 1) returning to pre-industrial food production, 2) imposition of strict government regulations and market requirements on farmers [6]. Rather, focus on agricultural residuals management of in-field and food processing biomass and wastage as renewable resources to sustain and improve soil productivity with focus on soil health and closing nutrients cycling gaps and reducing absolute dependance on external input [8]. I will add that CA does not mean abolition of synthetic chemicals use in agriculture, but, rather, responsible use to augment the natural system and increase agricultural productivity without posing a threat to the ecosystem and human health based on sound and science-based principles and practices.

The major principle of Circular Agriculture (CA) is based on; 1) minimal and responsible use of external inputs, 2) closing nutrients loops, 3) regenerating soils health 4) minimizing negative impact on the environment, 5) increasing biodiversity, 6) improving farmers' financial returns and livelihood, and 7) increasing resilience to climate change using practices that allows for circularity in the farming systems (Fig. 3 in Chapter 1) at different levels based on farmer capabilities and needs.

2.1 Minimal use of external inputs:

A circular agriculture seeks to use less inputs from outside the level of operation and to be more resilient in depending on its internal resources and partners [9]. For example, in a single farm CA where animals and crops are on the farm, the animal manure could help reduce external synthetic fertilizers brought to the farm (Fig. 3 and 4a). Also, the use of legume cover crops would help fix atmospheric nitrogen (Fig. 4b & c) and reduce external inputs. Also, the use of cover crops as catch crops (Fig. 4c) in off-season help to hold the residual nutrients in place, improves moisture holding in the soil, and improve soil health to benefit following crops reduces external inputs use on the farm.

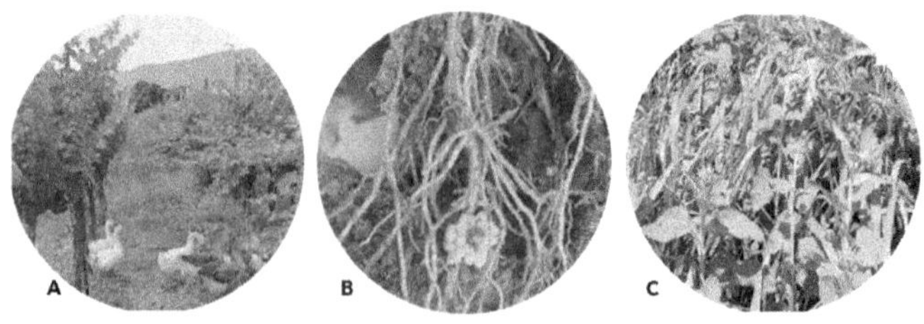

Figure 4: Integrated soil health and fertility management options for a circular agriculture systems. Geese integrated into the vineyard at Clear Creek Vineyard and Winery, Camp Verde Arizona (a), winter peas cover crop showing nitrogen fixing cluster nodules (b) and multispecies cover crops (c) at Whitestone Farm as a catch crop, Paulden, Arizona.

2.2 Closing nutrients loops:

Gaps that allow nutrient leakage on the farm such as leaching, runoff, and organic waste disposal to landfills can be closed using best management practices. Agriculture waste comes in different forms either on the farm, home, and processing plant with the majority ending up in the landfills (Fig. 2 in Chapter 1). It is important to know that these by-products that are called waste are resources if we invest time and resources to understand the by-product stream, chemical composition, and how to use them. For example, applying fertilizer in the right amount at the right time using the right methods would reduce leaching and runoff while implementing residue management, catch crops (Fig. 4b &c), and composting would keep excess nutrients and return some harvested nutrients to the field. In a CA system, closing nutrients loops equates using practices and technology that reduce the wasteful use of finite resources, promotes recycling and reuse of resources, and eliminate leakages (carbon, nitrogen, phosphorus, and water) in ways that enhances soil regeneration and an increase in biodiversity [1, 9,10].

2.3 Regenerating soil health and biodiversity increase:

Practices such as reduced tillage and conservation agriculture with focus on residue management as soil cover increases soil health and increased soil biology with additional benefits of reduced emissions. In a FAO report on the "State of the world's biodiversity for food and agriculture," biodiversity was identified and a key driver to increasing agricultural production and livelihood resilience against shocks such as climate change [11]. The current linear agricultural systems continue to destroy the biodiversity with the monolithic systems characterized by intensive and misuse of tillage practices and synthetic chemicals. These linear agricultural systems have impacted biological diversity in the soil, reduced soil health, reduced plant, insect, and animal diversity [1]. The use of regenerative agricultural practices such as organic amendments, rotations with cover crops ((Fig. 4b and 4c, Fig. 5), intercropping (Fig. 5), agroforestry systems (Fig. 4a), biofertilizers, reduced tillage, integrated pest management, pollinator habitats, etc. increase biodiversity, increase carbon sequestration, improves overall soil health for a resilient ecosystem that supports sustainable food production with minimal external inputs (Fig. 5). Also, regenerative agriculture reduces our dependence on cutting down fresh forest for more fertile and productive land required for agriculture productions, reducing the negative impact of agriculture on the environment.

Figure 5: Intercropping/multiple/mixed cropping system with diversified crop species (cabbage, clover cover crop, and wheat) on the same piece of land at the same time. (Credit to Alisha Cahlan, who took this photo at Braga Farm and Dole Windmills)

Also, there is and increasing interest in the use of biofertilizers as a regenerative practice to augment and reduce the synthetic fertilizers use. According to Data Bridge Market Research analyses, globally, biofertilizers' market size was valued at one billion United States dollars and projected to see a compounding growth rate of 12.7 percent annually between 2022 to 2029. Biofertilizers are beneficial microbes containing substances used to promote plant and tree growth through increased supply of essential nutrients. Mycorrhizal fungi, blue-green algae, and bacteria are the beneficial living microbial organisms used for these biofertilizers (refer to chapter 6 for more discussions on the role of biofertilizers in circular agriculture).

2.4 Minimizing negative impact on the environment:

Agriculture is a major contributor to deforestation, nitrogen leaching, and phosphorus and pesticide runoff that negatively affects the environment. Also, manufacturing and distribution of these inputs are related to greenhouse gas emissions. Circular agriculture has recycling, the use of internal resources and regenerative agricultural practices that help reduce direct and indirect negative impact of agriculture to the environment. For example, regenerative agriculture practices implementation, such as reduced tillage, cover crops (Fig. 4b and c), mixed farming systems with animal integration (Fig. 4a), and multiple cropping systems, would reduce the amount of easily leachable synthetic inputs on the farm, increase carbon sequestration, reduce soil erosion, reduce nitrate leaching into water bodies, phosphorus run off, improve biodiversity, and could reduce manufacturing and transport-related greenhouse gas emissions.

2.5 *Improving financial returns and quality of farmer's and community livelihood:*

When all the above six principles are well-established, the financial returns of the farmers would be realized, which would then improve their livelihood and resilience. Once farmer's livelihoods are improved by these circular agricultural practices, there would be resilience in food productions and supply of ingredients and raw materials for industry use. The focus should be on the quality and healthy food for human consumptions without health implications than the quantity [12-]. Though CA may be challenging in the early years of transition, the long-term benefits are enormous and will reduce cost associated to synthetic chemicals use burden that would increase their profitability, hence improve livelihood.

CHAPTER 3

The Role Of Circular Agriculture (Ca) Implementation Hierarchy-Effectiveness And Productivity

3.1.0 Circular agriculture at hierarchical levels of operation

Circular agriculture is not a one-size-fits-all approach to all farms and regions, due to diversity in farming operations in terms of resources availability and proximity, farming systems, culture, politics, and regulations. Hence, creating circular models at different sectors and levels of farms, communities, countries, economy, and culture with

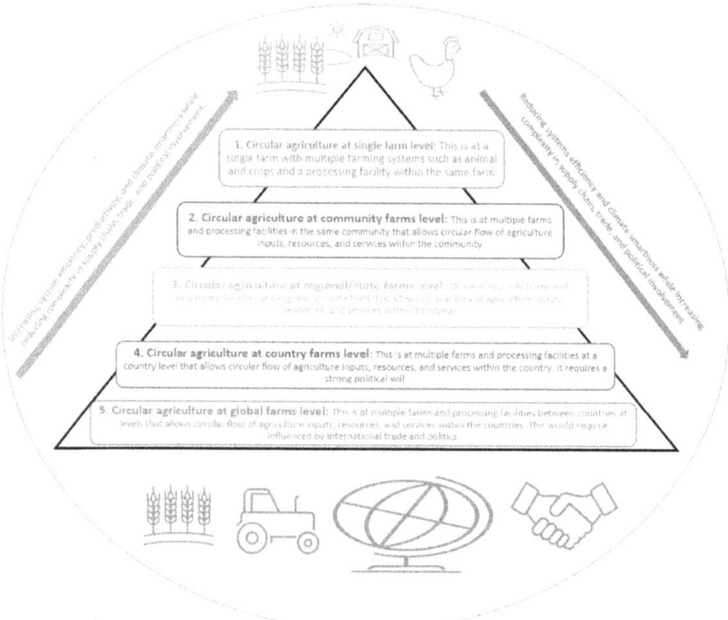

Figure 6: Proposed Mpanga's Circular agriculture implementation pyramid showing hierarchical levels of implementing circular agriculture, their productivity, efficiency, and the requirement for political and international trades involvement.

localization and understanding of local needs and resources could be a promising way to achieve sustainability [12]. Circular agriculture practices can be implemented at different hierarchical levels from a single farm to multiple farms at community, regional, and global space organized into what I call the Mpanga's circular agriculture implementation pyramid (Fig. 6).

Though these levels make it possible to implement CA at different levels, they vary in elements such as stakeholders' involvement and efficiency. Each level has a different number of farms involved, proximity of farming systems and processing facilities, efficiency and productivity, and political and international trade involvement (Fig. 6).

3.1.1 Circular agriculture at a single farm level:

This is at a single farm with multiple farming systems, such as animals, crops, and a processing facility within the same farm. It could be simple, with just crops and animal using all the principles of CA like agroforestry system with fruit trees and chicken (Fig. 7a) or a complex system that requires a processing facility (Fig. 7b).

At this level, the farm operation is at a stage where most inputs and systems required to be self- reliant are within the farm with less to no external inputs from outside the farm business (Fig. 7). For example, in Fig. 7b, the fishpond water used in irrigating the vineyard, the compost, and geese are all sources of nutrients to the plant without a need for external inputs. The farm has an on-farm processing facility (winery) where grapes are processed into wine and the grape pomace (by-product) from the winery composted and returned to the vineyard as fertilizer. Most of the wine is sold to customers visiting the farm and is the only product that is taken from the farm..

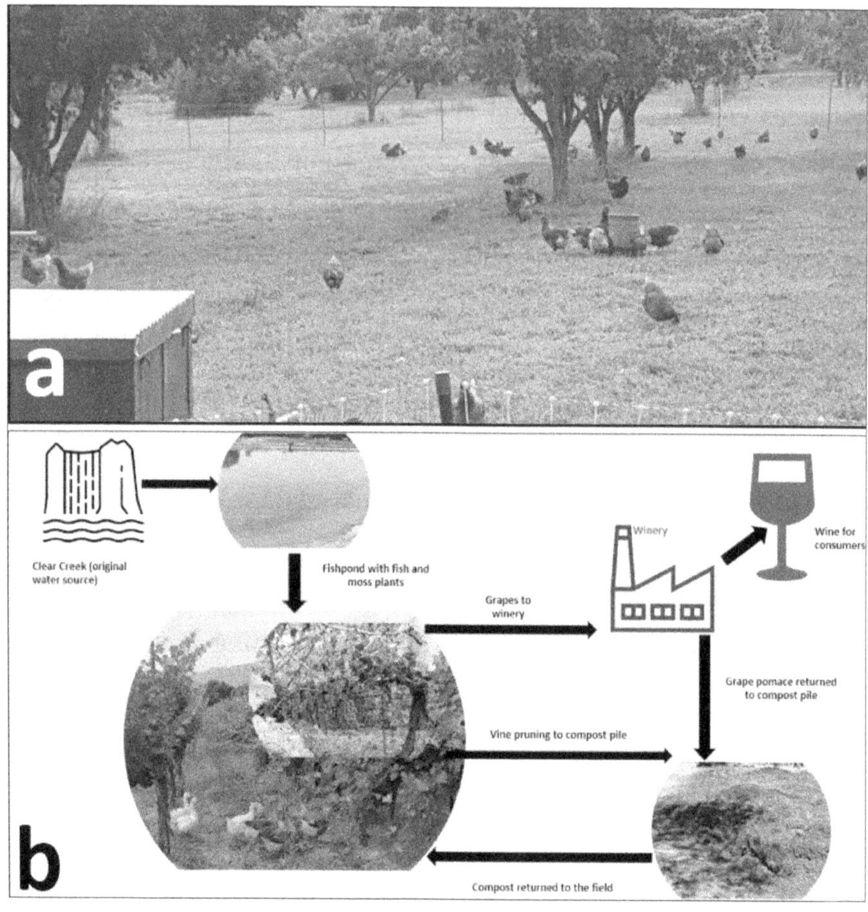

Figure 7: A circular agriculture (CA) model farm at a simple (a) and complex (b) single farm level CA implementation. a- has a simple system with orchard with well-maintained pasture that serve as feed for poultry and poultry dropping nourish the health of the land for both orchard and pasture. b- has a complex system with vineyard, geese, fishpond, winery, and composting facility.

3.1.2 Circular agriculture at community farms level:

This is at multiple farms and processing facilities in the same community that allows circular flow of agriculture inputs, resources, and services within the community. At this level, a single farm does not have all the necessary systems to make it self-reliant but rely on other farms and processing facilities in the community to complement themselves. For example, a vineyard that does not have a winery would supply their grapes to a nearby winery but make sure they take the grape pomace back into their compost pile. Also, the use of community recycled water on a farm within the community is a community level CA. In Cottonwood in Arizona, city recycled water is used to produce grapes and wine at Yavapai Community College, which is used to service tourist and community members interested in wine (Fig. 8 [15]).

Figure 8: *Yavapai Community College (Southwest Wine Centre) depends on Cottonwood City recycled water to produce grapes and wine that attracts tourist to the city (Mpanga, et. Al. 2022) [15].*

3.1.3 Circular agriculture at a regional/state level:

This is at multiple farms and processing facilities at a regional or state level that allows circular flow of agriculture inputs, resources, and services within the region. At this level of CA, high-level collaboration (among farmers and industry) and strong policies with political and social acceptance is required for success. These levels help tackle broader in scope environmental issues with potential funding. For example, regions and states can agree "no organic waste to landfill" policy, which means all organic waste must be returned to farms for composting and use as fertilizers (Fig. 9) or use in biodigesters for both energy and fertilizer production.

For example: A state like Arizona has one largest poultry farm called Hickman that uses the grains produced in the state to feed the chickens. The poultry farm also used waste plastic bottles to make their own packaging materials for the eggs. The eggs and meat from the poultry are all distributed and sold in Arizona. Also, the poultry manure from the poultry farm is processed into compost/organic fertilizers and sold back to the grain producers who then use it to produce grains (Fig. 9).

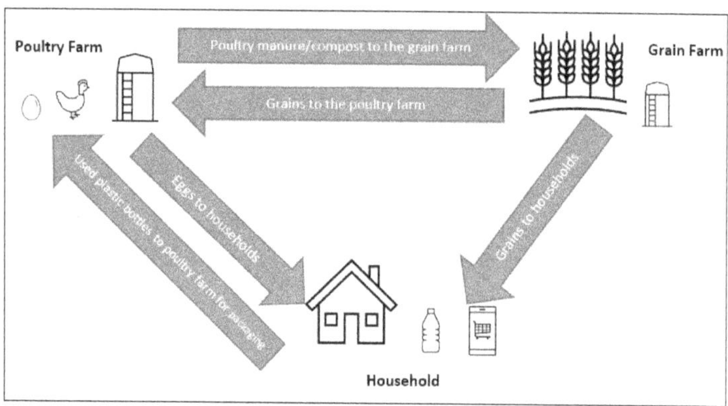

Figure 9: *A simple illustration of Circular Agriculture at the state or regional level where circularity exist between the largest poultry and grain farms and their communities across the state.*

3.1.4 Circular agriculture at country level:

This is at multiple farms and processing facilities at a country level that allows circular flow of agriculture inputs, resources, and services within the country. It requires a larger investment and strong political and social will. Examples, 1) a farmer has many farm locations in different states across United States that complement each other in resources use and help implement circularity on these farms. 2) Different farms and manufacturing plants in different states/provinces identified complementary resources and practices that help them implement circularity principles. It requires high-level partnerships with different sectors, including supply chain expertise and is mostly driven by political leaders and government policies.

The Netherlands would be a good example here as a country working on circular agriculture at the country level. In the Netherlands, Wageningen University is leading the research effort through stakeholder involvement towards CA. Farmer participation in policy and research with strong political will are key elements to CA success at this level. Using science-based research in a farmer and community participatory approach to make sure farmers accept the change. It must be planned with long-term slow adjustments with proper incentive to compensate for potential losses. Also, it is important to make room for flexibility using integrated management approaches.

3.1.5 Circular agriculture at global level:

This is at multiple farms and processing facilities between countries at levels that allows circular flow of agriculture inputs, resources, and services within the countries. This is strongly influenced by international trade and politics. This is much feasible at country borderlines, where complementary systems, such as factories and farms, exist across those border

3.2.0 Determining factors for CA level selection.

Currently, circular agriculture cannot be a one-model-fits-all situations because of the differences in farming systems, complexities in farming systems, distance to processing facilities, distribution of consumers (end users), proximity of other complementary farms, transport access, society priorities, political interest, and complexities in waste channels, etc. It is, therefore, especially important for anyone interested in circular agriculture systems to identify a specific level that fits best into their system as a starting point and work towards a more self-reliant and efficient level of CA system as demonstrated my Mpanga's implementation model in Fig. 6. The factors below would guide you to understand what level of CA is appropriate for you to start

1. *The goal for wanting to transition into CA*: What is your CA implementation goal? This is key to selecting the best systems and partners for your need. A financial goal to CA implementation (focus on more income) compared to climate-smart goal (focus on sustainable climate solutions) would both require different levels of CA implementation. For financial goals, all the five CA levels could help achieve it, however, only level 1 and 2 would be preferred options to help achieve climate-smart solution goals (Fig. 6).

2. *Existing agricultural systems, practices, and associate partners:* A good understanding of and taking stock of existing system are key factors to selecting the best fit CA level for your needs.
   Scenarios:
   a. A farm with diverse cropping systems, animal integration, and a processing facility of their produce can initiate a CA at the single farm level (Fig. 7) without overbearing challenges.
   b. A monoculture farm with only poultry production: starting CA at the single farm owner level may be too expensive and

complex and requires a big learning curve to succeed, hence starting at the community or regional level where partnership could exist with grain growers, consumers, and other partners as illustrated in Fig. 9 would be more feasible starting option.

3. *Economics and financial strength of both implementer and consumer:* What is the financial power of the farmer? What is the purchasing power of the end-user?

Scenarios:

   a. Moving from monoculture agriculture systems to Farm level CA would require noticeably big investments compared to transitioning to the other CA levels such as community, region, and country. This makes it more expensive to make such radical changes, hence, only farmers with great financial standing and/or the ability to produce at low cost can afford to make such momentous changes knowing that cost will not be a burden and/or constrain to the consumer patronage.

   b. Consumers who have good purchasing power and are willing to pay more gives a monoculture farmer a green light to succeed transitioning to any CA at level that is far apart his/her comfort zone. If the consumers have less purchasing power and would not be willing to pay more, then starting at a CA level with less investment is best option and then moving up the model with time.

4. *Consumer values and interest in sustainability and climate resilience:* The values and interest of the consumer in what you are doing as a farmer, manufacturer, a service provider, etc. is key to your success. Example: Consumers who value and have interest in sustainability would support you to move into CA, meaning that

reasonable investments to transition to CA at any level would likely succeed because pricing would not be issue. On the other side, consumers who do not value and have no interest in sustainability would not support you at any level, and you are likely to fail at any of the levels if it requires extra investment. Make sure you understand your consumers before deciding any CA transition plans starting at a level best fitting your business.

5. *Societal/community culture.* This point is like the consumer values and interest except that here, it is their way of life embedded in them, making change even more difficult. A CA at any level from 1 to 5 (Fig. 6) will succeed in communities and societies with pro-sustainability culture. Also, most CA levels are highly likely to fail in communities and societies with anti-sustainability culture if it requires extra investment.

6. *Governance and policy:* Political and institutional leaders, the priorities, values, and the policies they implement could influence CA implementation and their levels that make economics imperative. Some countries and their governance are more inclined to sustainability than others. Even in the same country, inclination to states and communities' governance and policies toward CA differs. Understanding local, national, and international governance and their policies towards CA would help identify best opportunities for best fit CA level selections to work with and improve with time. For example, governance with policies inclining to CA would potentially have incentives and support to growers which may offset initial cost to help them move faster to more sustainable levels such as the farm and community levels (Fig. 6) [2]. Also, friendly national policies and interest between

two countries would facilitate CA at the global level, especially among communities at the national borders. FAO reported that government across the globe policies are key drivers to influencing agricultural ecosystem services in different production systems [1].

7. *Science, technology, and research information availability:* Science, technology, and research information availability is a major driver on agriculture ecosystems services and practices. Farmers across the globe depend on science and technology for many decisions making on the farms (Fig. 10).

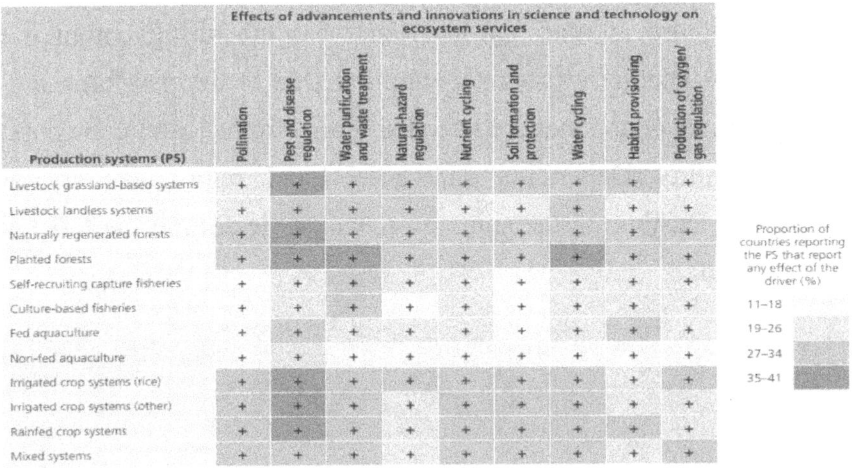

Production systems (PS)	Effects of advancements and innovations in science and technology on ecosystem services								
	Pollination	Pest and disease regulation	Water purification and waste treatment	Natural-hazard regulation	Nutrient cycling	Soil formation and protection	Water cycling	Habitat provisioning	Production of oxygen/ gas regulation
Livestock grassland-based systems	+	+	+	+	+	+	+	+	+
Livestock landless systems	+	+	+	+	+	+	+	+	+
Naturally regenerated forests	+	+	+	+	+	+	+	+	+
Planted forests	+	+	+	+	+	+	+	+	+
Self-recruiting capture fisheries	+	+	+	+	+	+	+	+	+
Culture-based fisheries	+	+	+	+	+	+	+	+	+
Fed aquaculture	+	+	+	+	+	+	+	+	+
Non-fed aquaculture	+	+	+	+	+	+	+	+	+
Irrigated crop systems (rice)	+	+	+	+	+	+	+	+	+
Irrigated crop systems (other)	+	+	+	+	+	+	+	+	+
Rainfed crop systems	+	+	+	+	+	+	+	+	+
Mixed systems	+	+	+	+	+	+	+	+	+

Proportion of countries reporting the PS that report any effect of the driver (%)

11–18

19–26

27–34

35–41

Figure 10: Government policies as driver to innovation in sciences and technology supporting ecosystem services by production systems around the world (91 countries) (Sources: Food and Agriculture Organization of the United Nations. Reproduced with permission.) [1].

Same applies to selecting CA levels—the more applicable technology availability with science-based research information speeds up innovation acceptance and implementation. Make sure to consider information and technology availability on practices, systems, and partnerships available for support. The more readily available science-based information and technology, the more likely you are on the success path. Farmers' involvement at all levels of CA technology and research discovery, trials, and implementations through farmer-participatory research and innovation approach is a catalyst for innovation and research information availability and acceptance by stakeholders.

8. *Availability and proximity of other farm and manufacturing:* Apart from the CA at single farm level, which has all components from production to manufacturing without selling any raw products except the finished product to consumers, all the other CA levels required partnerships either with other farmers or with manufacturing companies. For example, a sole indoor fish farmer will require a grain producer and manufacturing company to partner. The grain producer supplied feed to the fish farm and is dependent on the manure from fish farm (from cleaning the ponds) for fertility. Depending on where (local, region, country, and globe) feasible partnerships exist would determine the CA level to practice.

CHAPTER 4

Types of Circular Agriculture Systems and Implementation Using The Hierarchical Model

Circular agriculture can be grouped into two basic types of systems based on whether they are in the open or indoor. These groupings are, in most cases, dependent on other CA types and work best in integration, rather than practicing them in sole isolation. The types of CA systems mostly have the primary and auxiliary systems. The primary system is the one that produces the actual produce/product of interest with auxiliary systems that make it possible for the primary to function as a circularity agriculture. The main CA systems are 1) open field circular agriculture 2) indoor circular agriculture systems.

1. Open field circular agriculture system: The produce and products from this system are produced from the open field agricultural systems such as grains, fruit trees, aquatic, animal husbandry or a combination of two or more. The combinations of animals and crops make CA practices more practical than just crop or animal. In a combined system, there is always primary and auxiliary products, which depends on the partners and the level of CA practiced. The primary products in this system could be from crops, aquaculture, and animals or more than one product. The primary products obtained from this system by a single farmer could be single or multiple, depending on the level of CA practiced.

Examples of open field circular agriculture at different levels:

a. *At a* single farm CA level 1 (Fig. 11) where one farmer owns the vineyard, poultry, aquaculture, and processing vine plant, the primary products from this system would be four (the grapes, the fish and poultry, wine, and animal processed product) (Fig. 10).

Figure 11: Open field circular agricultural system with grapes, poultry, and aquaculture, at a single farm circular agriculture.

b. At community, regional, country, and global CA levels 2-5 where different farmers and partners own different components (the grain farmer, the animal farmer, the grain processing plant, and the meat processing plant) in Fig. 12, each partner has a single, primary product and other partners' products becomes auxiliary. In this case, the grain farmer only has grain as his/her primary product while all other components become auxiliary products that support him/her to implement CA in the context of the community, region, country, or globe.

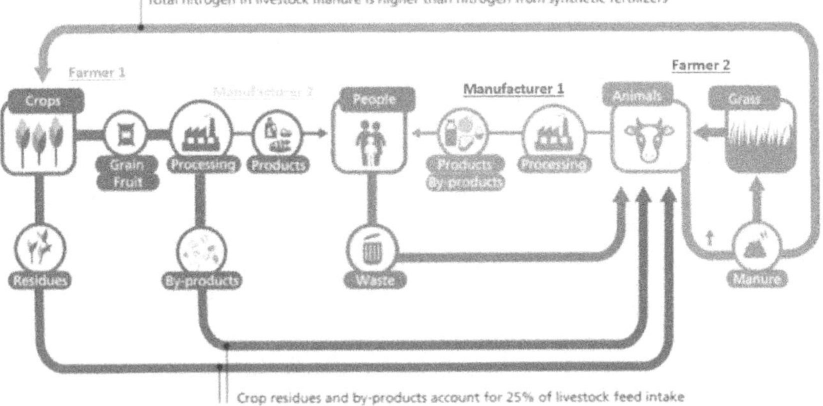

Figure 12: Open field circular agricultural system with livestock, grain, and a processing plant at community, regional, or country level of circularity. The arrows are connecting different components with long travel distances from one another (Adopted: Food and Agriculture Organization of the United Nations. Reproduced with permission.) [12]

2. Indoor circular agricultural systems: These systems are mostly in confined areas with full to no climate control systems that help reduce the impact of outside weather conditions. They require less land but have high initial capital and maintenance requirements. Example will be greenhouse productions. The primary and auxiliary produces, and products obtained from these indoor CA could be crops, animal, and aquaculture. This could also be sub-classified into 1) circular horticulture, 2) circular indoor animal husbandry, and 3) circular indoor aquaculture, or 4) a combination of 1, 2, and 3.

Examples of indoor circular agriculture system at different levels.

a. At a single farm level of CA where all the components are located on a same farm and managed by the same team on a highly closed approach. In fig. 13, fish is produced in tanks, the water with fish fecal matter, which is a source of plant nutrients are used to irrigate crops. The plants utilize the mineral and makes the water clean for use in the fishpond to produce fish.

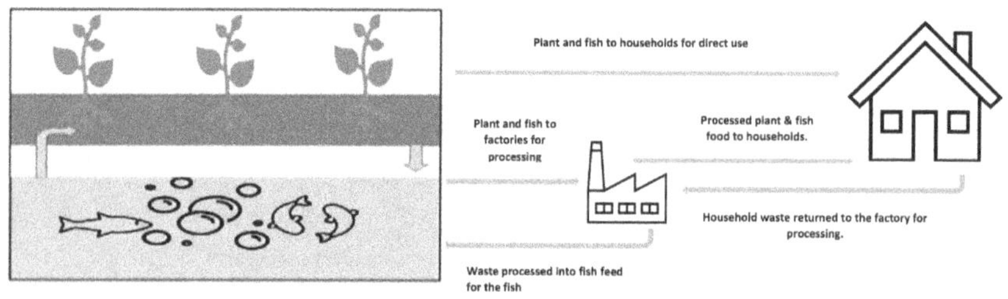

Figure 13: *Indoor circular aquaculture aquaponic system at a single farm level (same location and same owner of the farm and processing plant).*

The most important unique characteristics at a single farm level of indoor circular agriculture is that all the systems (plants, fish, and processing) are indoor on the same farm and is owned by a single farmer/business. The only factor outside the system is the consumer or the households that uses the products. (Fig. 13). It has the advantage of full control, efficient in recycling, and reducing greenhouse gas emissions due to very limited transportation.

b. At a community/regional/country farm level of CA where all the components are long distances away from each other on different farms and managed by different teams. The most important unique characteristic of indoor circular agriculture is that all the systems (the farm and the processing) are far apart and in various locations (Fig. 14). Also, at this level, more transportation is involved for the movement of materials.

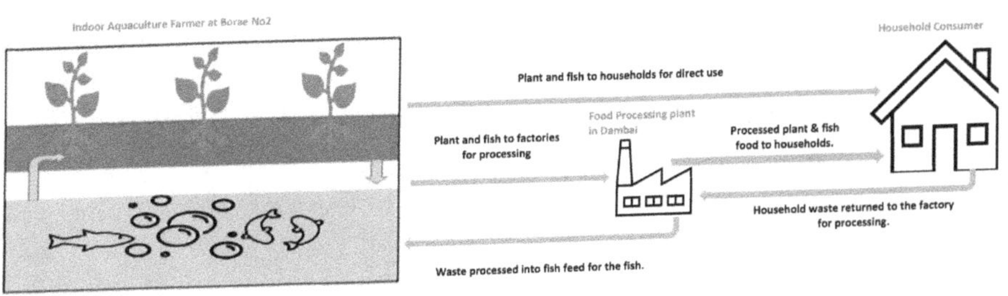

Figure 14: *Indoor circular aquaculture aquaponic system at a community, region, or country level (separate locations and owners of the farm and processing plant).*

Comparison between open field and indoor circular agricultural systems

The open fields and indoor circular agricultural system can be practices at different levels of circularity in agriculture ranging from a single farm to country level. Their level of complexity, resources required, profitability, and land requirements could differ widely, as illustrated in Table 2.

Table 2: comparison between open field and indoor circular agricultural systems

<u>Comparison factors</u>	Open field circular agriculture	Indoor circular agriculture
Initial cost	**	***
Heavy duty machinery needs	***	*
The use of fuel	***	*
The use of electricity	*	***
Greenhouse gas emission	***	*
Carbon sequestration	***	*
Biodiversity	***	*
Technical knowledge requirements	**	***
Profitability	**	***
Intensive care requirements	*	***
Land area requirements	***	*
Water requirements	**	*
Integration into urban planning	*	***

*low   **=medium   ***high

CHAPTER 5

The Role of Food Waste Management in Circular Agriculture and Economy

The United Nation Environmental Program (UNEP) Food Waste Index Report 2021 estimated about 931 million tons of yearly global food wasted. The report also suggested about eight to ten percent of global carbon emissions are associated to an estimated 17 percent of food that may go wasted around the globe with 61 percent coming from households, 26 percent from food service, and 13 percent from retail [13]. According to an EPA 2020 report on food waste, 103 million tons food waste was recorded in the U.S. in 2018 (Table 1 and Fig. 2) [5]. Other organic waste materials exist, such as yard waste and on-farm waste (manure and plant residuals) that unsustainably ends up at the wrong places such as landfills, which courses more environmental burden on our ecosystems. The UN Sustainable Development Goal 12.3 has a key focus on reducing food waste at all levels [13], which includes reuse into other products, such as compost, that helps return the resources into the field for food production. In the United States, great efforts are made to divert these food waste into other sectors of the economy (animal feed, biochemical processing, anaerobic digestion, composting, land applications, etc.) (Fig. 15) that make excellent use of the resource. However, more needs to be done to ensure none of these materials are wasted.

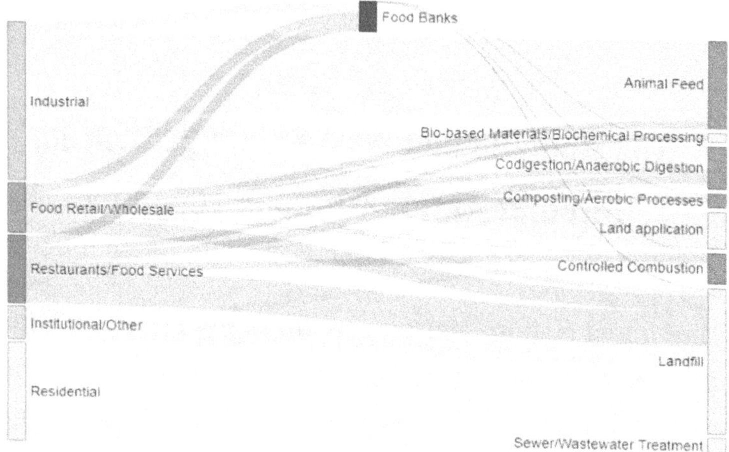

Figure 15: *Sources of food waste and management channels in the United States in 2018 (EPA, 2018 Food Waste Report) [5]*

5.1 The role of composting in circular agriculture

Composting is an old practice used to turn organic waste into extremely resourceful fertilizer (Fig. 15 and b) in for both backyard gardening and commercial crop production across the globe, due to its great benefits relating to soil health, low cost, and a sustainable waste management strategy, which is the focal discussion in this chapter—the role of compost in circular agriculture. The EPA food waste report for 2018 indicated that only 2.6 million tons of food scraps were composted from the 35 million tons that went into landfills [5]. More investments are needed to turn this important resource into fertilizer through composting that will go into farmlands and reduce the burdens on landfills.

5.2 What is composting?

Composting is commonly done in a controlled environment [(containers, pits, and covering with aerobic (oxygen-required)] that help transform organic waste materials through decomposition with the help of microorganisms into soil amendment or mulch with nutrients that are utilized by plants. Properties of a matured compost include dark (Fig. 15a and b), crumbly, earthy-smelling material. The microorganisms break down the materials using carbon and nitrogen for nourishment and reproduction, water supports materials digestion, and oxygen for breathing.

Figure 16: On-farm ready-to-use grape pomace composted with fish waste at Page Spring vineyard in Arizona (a) and commercial city level City of Denton Dyno Dirt composting site in Texas ready to use compost (b).

5.3 Composting methods:

Composting can be done in the backyard, on the farm (Fig 16a) and in commercial scale (Fig 16b) using any of the methods below.

1. **Aerobic composting:** This is the most used composting method with regular turning of the pile to allow oxygen exchange.
2. **Anaerobic composting:** Here, the composting is done in an air-tight container or space without oxygen and no turning of pile.
3. **Vermicomposting:** is a form of composting approach with worms (earthworms) added to the compost pile to speed up breakdown and decomposition or organic materials. The worms are selected based on their ability to feed and breakdown specific organic materials.

5.4 How to compost using aerobic composting

Composting Ingredients: Four key factors (nitrogen, carbon, air, and water) are needed for microorganisms to function effectively in breaking down compostable organic material (containing carbon, with varying amounts of nitrogen). For composting to be successful, the right ratio of carbon to nitrogen (30:1) must be maintained, as well as the right amounts of air (by turning regularly but not too frequently—every two weeks is recommended—and water (draining when the pile becomes too wet and spraying mist of water when it becomes too dry).

Note: Too much carbon-rich material (dry and brown organic material) makes your compost pile dry, hence, it takes longer for your compost to mature. Also, excess nitrogen-rich material (green plant materials and kitchen waste) could make your compost pile slimy, wet, and smelly. The best rule of thumb to achieving the right carbon to nitrogen ratio (30:1) is to add four to three parts of carbon source material (brown) to every one part of nitrogen material (green). Also, it's important to note that warmer temperatures speed up decomposition more than cold temperatures, due to active microbial activities in warm temperatures. To ensure mature compost is free of pathogens and weed seeds, regulations by the United States Environmental Protection Agency specify that compost should maintain a minimum operating condition of 40°C for five days, with temperatures exceeding 55°C for at least four hours within this period. It's important to regulate temperatures from exceeding 60-65°C (most beneficial microorganisms cannot survive above this temperature range) by turning or aerating the compost pile regularly.

Importance of composting in a circular agriculture and economy.
Composting help;
- Keep food waste out of landfills and make them useful while keeping the environment clean.
- Reduce greenhouse gas emissions associated with producing and transporting food and not using it.
- Use finished compost in a garden to recycle nutrients back into the soil to improve soil health.
- Expand the soil's ability to store carbon through the improvement of soil health.
- Reduce the need for synthetic fertilizers and pesticides inputs and the associated emissions.
- Reduces pathogen risk associated with direct use of some waste, such as animal manure.
- Reduce soil erosion through improved soil water infiltration and water storage.

- Increase farmers' livelihood through increased yield with low-cost compost inputs.
- It is a source of employment for those collecting the waste, composting, and distributing.
- Adds monetary value to economies directly (sales and use of compost) and indirectly (increasing resilience to climate variability)

5.5 A case study on how composting can drive circularity at the city level

Dyno Dirt compost is the City of Denton in Texas's own composting facility using city organic waste from yard trimmings and recycled biosolids. The city of Denton in Texas, United States of America and the area residents and professionals have been using Dyno Dirt products since 1997 and have benefited from it for most backyard gardening (Fig. 17), landscape planting and mulching, lawn establishment and maintenance, flower gardens, nursery crop and green house production, and turf and sod production.

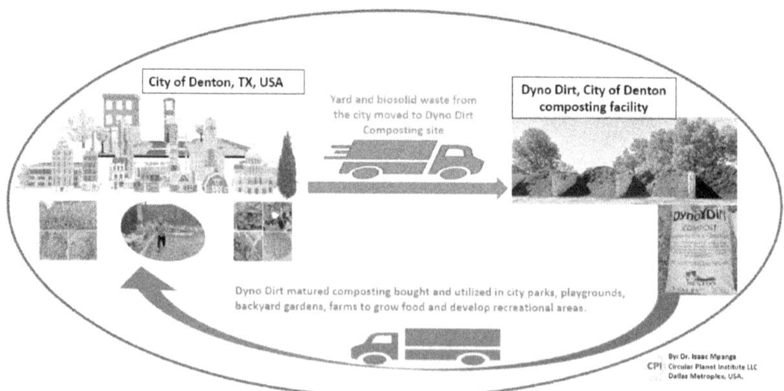

Figure 17: City of Denton's Dyno Dirt composting program showing the role of composting in circular agriculture and economy.

The compost has shown high and consistent nutritional composition with low heavy metals and microbial pathogens from 2020 to 2022 laboratory testing they have been doing. The results from the Dyno Dirt Compost shows that most organic waste can be composted with maximum benefits and must be composted using the right protocols from Environmental Protection Agency. The compost had about 11 kg total N per ton and 9 kg organic N per ton with 5 kg P pe ton. The City of Denton composting program is an example on how composting can be used to create value in circular economies through implementation of circular agriculture at the community level (Fig. 6) that meets CA principles (closing nutrient leakages, turning waste into resources, increasing soil health and fertility with less dependance on external synthetic inputs, in the city, potential emissions reduction from reduced associated chemical fertilizers use, increasing environmental health and livelihood for the citizens) (Fig. 17). Fig. 16 is a good example with strong indication of the key role of composting in circular agriculture and economies at a community and city level CA implementation as a climate-smart tool.

5.6 Challenges associated to using compost in agriculture.

1. It is labor-intensive for large-scale production that requires more land and space for operations.
2. It requires specialized equipment, such and the turner and mixers on larger scale production.
3. It's bulky to transport from one place to another with less concentration of minerals.
4. It could be a source of environmental pollution if the right protocols are not followed for composting. Even with good

composting, related issues such as microplastics cannot be removed just by composting. Microplastics can only be avoided in compost when plastic pollution is eliminated from composting materials. In the case of using biosolids, special treatment needs to be carried out to remove the microplastics before composting.

5. If not well composted, it could be a source of pathogens to both plants and human.

6. Limited data on how to effectively collect household organic waste without contamination with unwanted materials

CHAPTER 6

The Role of Biofertilizers in Circular Agriculture and Eeconomy

Intensive use of agrochemicals has been the focus to improving agricultural productivity to provide food and shelter for the increasing population. This is characterized by overuse of synthetic nitrogen (N), phosphate (P) fertilizers, and pesticides, which have environmental-related issues to groundwater pollution, eutrophication of surface waters, and increasing greenhouse gas emissions. Overmining of limited rock phosphate, which is concentrated in a few locations around the globe due to extreme rates of P fertilization led to high-risk perturbation of the P cycle. The Challenge with P is that 99 percent of the total P fractions (either applied or already in the soil) are strongly limited in availability for root uptake due to fixation properties to Ca (in alkaline soils), Al, Fe (in acidic soils) and may need special mechanisms by the plant to access it. Other related factors, such as crop development stress, impairs root development, with increasing impact related to climate change, which calls for more innovative and strategic approaches to recycling nutrients, closing the leakages, and improving efficiencies of N and P. Biofertilizers which are microbial organisms and products isolated and enhanced for soil health improvement have been identified as one of such innovations with fast increasing market share globally. These biofertilizers have the potential to strategically decrease synthetic N and P fertilizers in agroecosystems by enhancing nutrient use efficiencies of fertilizers based on products of waste recycling, right timing, and placement of fertilizers, and biofertilizers with plant growth-promoting microorganisms. Market predictions indicates that the biofertilizer market will register 14 percent growth by the end of 2023. As of 2016, the biofertilizer market size was USD 1106.4 million globally with projections to grow at 14.2 percent rate to reach USD 3124.5 million by the end of 2024 [14].

6. 1 Fertilization strategies with biofertilizers to increase efficiencies of fertilizers

Plant growth-promoting microorganisms (PGPMs) and products use as biofertilizers is gaining interest for many reasons relating to environment and human health and the rising cost of synthetic fertilizers. PGPMs and products could be integrated into nutrient management in an integrated nutrients management approach where they are used with other forms of fertilizers to increase their efficiency through induced root growth promotion, organic acids release, mineralization of organic matter, production of volatile compounds, etc. [16]. This section discussed how PGPMs could be used strategically to support nutrient acquisition of crops through improving fertilizer use efficiencies to enable lesser fertilizers' use for crop productivity, and to reduce detrimental environmental side effects related with high inputs fertilizers.

6.2 Case studies on how PGPMs increases organic fertilizers efficiencies

An integrated nutrient management approach combining compatible fertilizers sources such as rock phosphate, compost, manure, and the right amount of synthetic fertilizers offers the potential management option to improve the performance of PGPM-assisted production strategies with improved efficiencies of fertilizers, especially organic sources for plant performance. In a range of experiments in maize, it was showed that sparingly soluble Ca-phosphates could be improved synergistically through the combination of PGPM inoculants with stabilized ammonium fertilizers (Fig. 18) [16]. The PGPM-ammonium combinations with sparingly soluble Ca-P supply reached about 84 percent of the shoot biomass production and 80 percent of the shoot P accumulation, as compared with positive controls fertilized with soluble P [16,17].

Figure 18: Combine application of stabilized ammonium and plant growth-promoting microorganisms improved rock phosphate efficiencies and maize performance comparable to positive control with soluble. P [16,17] (Picture by Dr. Isaac Mpanga at the University of Hohenheim, Stuttgart, Germany)

In another approach, the combination of PGPMs with organic waste recycling fertilizers, such as municipal waste compost or composted poultry manure (PM compost), manure, etc. on low P soils with contrasting pH in Ghana showed improved efficiencies of the organic fertilizers [18]. Also, this was confirmed in a meta-analysis under the European Union's BIOefFECTOR project with improved efficiencies of organic-based fertilizers such as manure, rock phosphate, and compost (Fig. 19a). Also, large-scale greenhouse and open-field tomato production trials conducted in Romania and Hungary revealed reproducible effects on yield and fruit quality over three years by PGPM combinations with manure-based fertilizers [19].

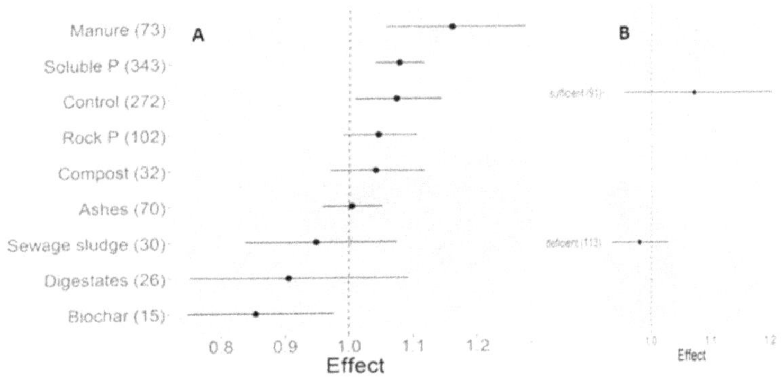

Figure 19. Effect of BE applications as a function of the type of P fertilizer added in the experiment (A) or the level of soil P availability (B) on either, grain, fruit, or shoot dry matter. The number inside the brackets represents the number of observations included the dashed vertical line indicates no difference between BE and non-inoculated control treatment; the points indicate the mean effect while the horizontal line represents the 95 percent confidence interval (CI). If CI lines cross the vertical line, effects are not significant. (Source: European Union BIOefFECTOR project, 2017) [19]

Final report of the European Union BIOefFECTOR project that was aimed at resources preservation in their agricultural systems [19] showed in the project meta-analysis that biofertilizer seem more;

1. Promising in controlled environments and nurseries production systems when compared to field productions [19]

2. Effective when combined with organic-based fertilizers (Fig. 19a) and ammonium sources in an integrated approach than sole application of the biofertilizers (Fig. 18) [16-19].

3. Effective under stress conditions when biofertilizers has a mixture of strains compared to single strains of bacteria and fungus. Interestingly, in non-stressful application conditions, mix strains and single showed the same effectiveness on crop yields.

4. Effective and improved yields and efficiencies of organic P sources such as manure, compost, sewage sludge, ashes, and rock P, except digestate and biochar (Fig. 19a).

5. Effective in pH range of 5.5 to 8.5 with more prominent efficiencies on a crop yield at neutral to 8.5 [18,19].

6. Effective at low substrate organic carbon with higher yield at 0 to 1.5 percent substrate/media organic carbon

7. Effective and improved crop yield when soil P was either low or moderate with less effects on high P substrates/soils (Fig. 19b)

6.3 Perspectives and opportunities of biofertilizers use in circular agricultural systems

Biofertilizers will continue to play a key role in circular agriculture and economy to enhance agricultural resources' use efficiencies and reducing risk associated to the sole dependency on chemicals. Associated benefits of biofertilizers includes biological nitrogen fixation, phosphorus solubilization from organic sources (Fig. 19a) [18,19], minimize stress relating to diseases and drought, and mobilization of other nutrients, including micronutrients, which supports yield increase, efficiencies in resources' use, reduce cost of inputs, recycle waste into resources, improve efficiencies of organic based fertilizers, and more (Fig. 19a & b) [16, 18-20] that could lead to improved livelihood. However, their efficiencies are still biased due to limited reproducibility of the expected effects under different production conditions, which is attributable to sensitivity of plant-biofertilizers interactions to environmental stress factors, particularly during the phase of establishment and to limited knowledge on positive or negative interactions with the native soil microbiome and the application conditions required for successful rhizosphere colonization (a pre-requisite for beneficial plant PGPM interactions). Also, soil pH-buffering capacity, particularly high alkaline soils and high organic matter and soil P were identified as limiting factors, counteracting the plant growth-promoting potential of selected inoculants with a proven ability for Ca-P solubilization and their overall performance on yield [19].

More commercial-scale localized field trials are required in various locations to ascertain efficacy of specific biofertilizer products with more collaborations in academia, industry, and farmers. To use biofertilizers in areas with limited localized data on specific products and strains, start with small pilots to assess their potential before expanding to larger fields with bigger investments based on their applicability to your location. Industries involved in the production of these products should involve university researchers for a better understanding on their mechanisms and growers for better understanding and ownership by farmers based on their locally prevailing conditions.

CHAPTER 7

The Role of Solar Farming (Agrivoltaic) in Circular Agriculture

The potential of solar photovoltaic was a great discovery for green energy until the competition with farmland for commercial scale met strong resistance from farmers, food activists, communities, and many more agricultural enthusiasts who have legitimate concern it will lead to food crises in the future due to farmland lost to solar energy production. This resistance from sole use of farmlands for solar energy production led to an integrated and more conservative approach called agrivoltaic (a deliberate co-existing of solar energy and farm on the same piece of land) for dual benefit of solar energy and food production [20-24]. This deliberate integration of agriculture and solar energy production is aimed at alleviating land use competition for the two economic sector [20] and to boost revenues for landowners [20], and other benefits that includes environmental and social. Innovatively, agrivoltaic systems have demonstrated to be technically and economically practical for a more conservative agricultural land use and have the potential to integrate food and energy production with estimated land productivity increase by 35 to 73 percent (Fig. 20) [21].

Figure 20: *Agrivoltaic design adapted from the United States Department of Energy project called Innovative Solar Practices Integrated with Rural Economies and Ecosystems (InSPIRE)*

7. 1 Types of agrivoltaic systems

The types of agrivoltaic systems are based on crops, animal, and aquatic agricultural systems integration with solar energy production.

7.1.1 Agrivltaic with plant cultivation:

This is the integration of plants and crops with solar energy production. The design could come in different forms such as using border fields, in-fields, and above plants (air space and roof in the case of greenhouses) (Fig. 21 a-c).

a. The in-field and border design (Fig. 21a) uses any part of the farmland for solar panels installation with compatible crops. More energy can be produced from the in-field and border inclusion design but may limit crops equipment use and limit space used for food production.

b. The only border design (Fig. 21b) is set for the solar panels for energy production to be only on marginal lands such as the borders without crops. This design allowed for more food production than solar energy and can accommodate broader crops and equipment use.

c. The air space/roof design of agrivoltaic (Fig 21c) uses the space above the crops in the field or the roof in the case of a greenhouse for solar energy while producing food on the same land. It is becoming common to have solar panels on a greenhouse roof for most controlled indoor producers as their source of energy for the farm. In the field with more shade from the panels, crop selection must consider shade tolerance/loving plants and not too tall plant species, such as shrubs and plants that can be trained to grow below the panels. Equipment use would also be limited due to the numerous panels in the field.

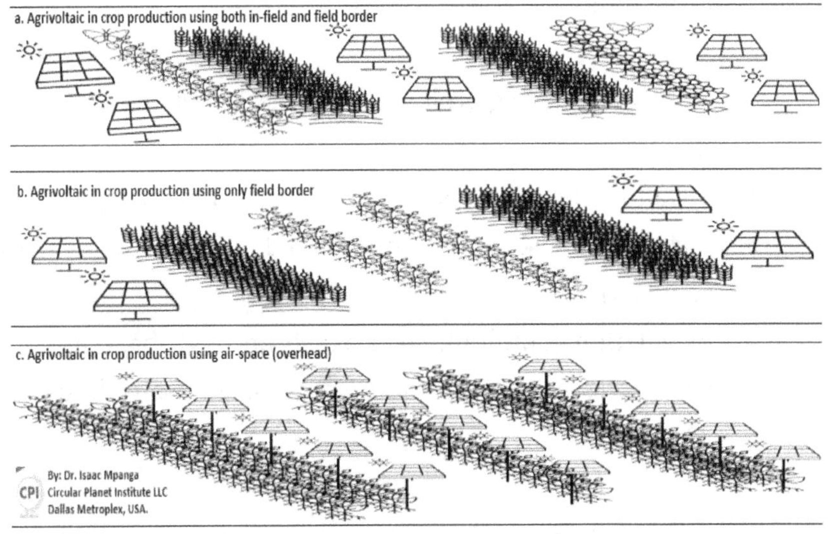

Figure 21: *Forms of agrivoltaic designs for solar energy and crop production integration on the same field at the same time.*

7.1.2 Agrivoltaic with aquaponics:

This is the integration of solar energy production into aquaculture systems (Fig. 22a-c) at the same time in the same body of water to produce both energy and food on the same farm. The main benefits of this system are;

a.   reduce evaporation from water surface.

b.   increase income stream for farmers through diversified souces.

c.   produce green energy doe the farm there by reducing energy related emissions.

d.   reduce energy cost on the farm thereby increasing farm profitability.

The design of agrivoltaic with aquatic systems could come in different forms such as

a.   installing the solar systems in the water body (Fig 22a)

b.   installing the solar system at the edges of the water bodies (Fig 22b)

c.   installation of multiple solar panels on single bars supported with poles erected in and/or outside the water body (Fig. 22c).

The selection of design must be based on farm objectives, capital availability, materials, expertise, the type of aquaculture (only fish, only plants, and both fish and plants), and species tolerance to shading. To maximize the sunlight interceptions irrespective of the design, integrating sensors to rotation panels in the direction of the sun are recommended.

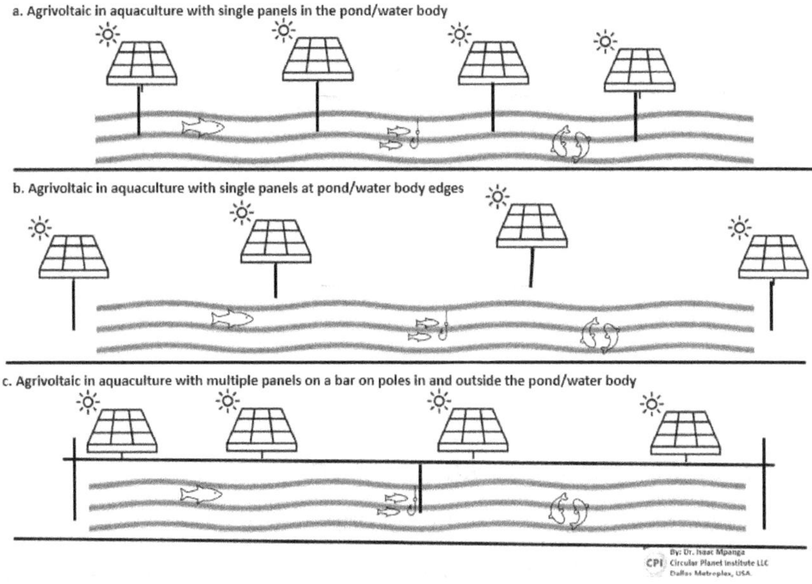

Figure 22: Different agrivoltaic designs in aquaculture systems.

7.1.3 Agrivoltaic with livestock production:

This system combines solar energy production with animal farming systems (mostly in relation to pasture management systems) (Fig. 23). Here, there are mutual benefits to the animal (shading and water harvesting for the animal to drink) and solar panels installed (animals eat the vegetation to prevent them from growing over the panels). Animals should be carefully selected to avoid aggressive animals that could climb/attempt to climb and cause damage the panels. Also, the grasses and the animals must be shade-loving species. For example, climbing or crippling plants may not be a good fit, since they could take over the panels if not controlled frequently and prevent good sun interception. Examples of good animals to use are sheep, hornless cattle, chicken, ducks, geese, turkey, etc.

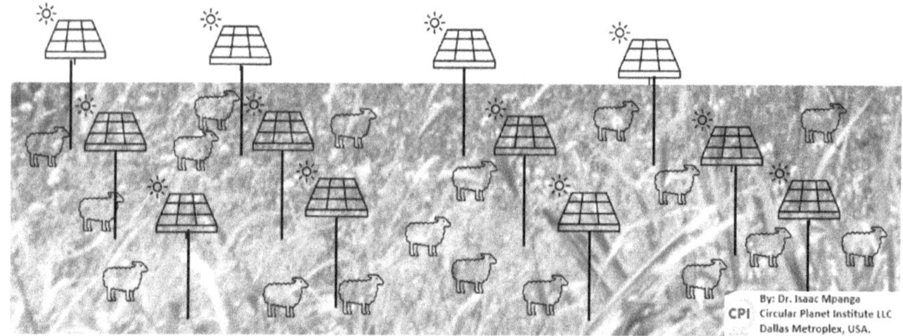

Figure 23: *Agrivoltaic with animal farming showing integration of solar energy production and sheep pasture*

7.2 General benefits of agrivoltaic systems

Agrivoltaic systems could come with many gains to solar developers, the farmer, environment, and the general society/community.

7.2.1 Benefits to solar developers are

1. Reduction in installation costs – using previously tilled agricultural farmland could exclude the need for ground leveling before installation of the solar panels.
2. Reduced upfront associated risk that comes with geotechnical risk assessment of non-agricultural lands. Previously tilled agricultural lands are identified as "least risk option."
3. Reduced legal risk during the environmental review associated to the use of non-arable land. Installing solar panels on previously disturbed land compared has low legal fees compared to high legal up-front risk associated with non-agricultural lands like virgin forest.
4. Potential increase in solar performance – The vegetation

under solar modules could lower soil temperatures around the modules especially on extremely hot summer days, hence increase solar performance.

7.2.2 Benefits of agrivoltaic to agricultural land managers are:

1. Reduced electricity costs.
2. Reduced greenhouse gas emissions associated other non-renewable energy sources on the farm.
3. Diversification of the revenue stream to the farmers.
4. Increased ability to install high-value, shade- resistant crops for new markets.
5. New market opportunity to sustainability-mindful audience and customers.
6. Ability to maintain crop production during solar generation with potential water saving and extended season.
7. The potential and luxury to allow degraded land to recharge or increase crop rotation years due to more income streams.
8. Increasing biodiversity.
9. Water-saving and use reduction potential due to shading from the solar panels that could reduce evaporation and the ability to harvest rain for targeted crop watering.
10. Seasonal extension potential due to the shading from the extreme sunlight in the summer months and protection from direct cold in spring and fall months. This allows for more time for farming activities.

7.2.3 Benefits of agrivoltaic to environment:
1. Reductions in energy-related greenhouse gas emissions associated with non-renewable sources.
2. Reduction in deforestation and the exploitation of virgin lands either for solar energy or new farms.
3. Increasing biodiversity through better micro-climate and more crop rotation years.

7.2.4 Benefits of agrivoltaic to society/community:
1. Increase energy security with clean source of solar power.
2. Potential water saving in agriculture, especially in warmer and dry regions.
3. Employment to the communities for both energy and farm industries.
4. Integration of energy and food into one system to make communities more resilient in both food and energy supply.

7.3 What role does agrivoltaic play in circular agriculture?

1. Energy consumption in the form of fuel and electricity are a major part of farm operation cost externally sourced outside the farm and are large contributors of greenhouse gas emissions. The ability of a farm to generate its own power will eliminate the external factor and make the farms more circular in producing their own energy at the farm and community levels of CA where they can share their green farm power with other farms.
2. The additional and diversified income from the solar energy production make the farms more productive, hence, the

ability to implement other CA practices that take many years to break even and make profit without hurting their operations. In other words, it gives insurance and security to the growers to try new innovations that are CA practices.

3. Increased natural resources conservation, such as water and land use.
4. Increase biodiversity in the cases where pollinator habitats are established with the solar systems.
5. Increase resilience to climate change with sustainable food and energy production for the increasing population.

7.4 Opportunity to scale agrivoltaic as a CA practice

1. The solar and energy industry, including policymakers, must work harder to gain the trust and acceptance of communities, especially farmers and food advocates, that agrivoltaic is a solution to the issue of farmland grabbing for sole solar installation, which was a threat to food security.
2. Farmers and communities must be drivers leading the initiatives and innovations of agrivoltaic, including decision-making based on their local conditions.
3. Transparency and sharing of information, learning, and opportunities for open dialogue are key to get stakeholders to buy in and have ownership.
4. To make progress on gaining acceptance and trust that will help establish integration of energy and food, the benefits of agrivoltaic to the farmer, the community, and the environment must be the top priority, rather than profit to coorperations.

CHAPTER 8

The Role of Integrated Pest Management in Circular Agriculture

Managing pest and diseases is a vital component to agriculture productivity, profitability, and food security around the globe. Plant pests, weeds, and pathogens have devastating effects on agriculture with a negative impact on food security and is estimated to potentially reduce production of food by 20–40 percent [25]. Total loss of food production is projected to be U.S. $40 billion per year worldwide, due to pest and diseases [25]. Unfortunately, agrochemical applications are the main approaches to controlling these plant pests and diseases, especially among the large-scale producers. According to the Meghmani 2023-2028 crop protection report, global crop protection chemicals' market trend has seen a growing trend (projected at four percent increase from 2023 to 2028 with over U.S. $70 billion by 2028) and will continue to grow due to increasing pressure on food production from human population growth and climate change. A study by Naidu, et. al. revealed that 64 percent of global agricultural arable lands are at risk of pesticides pollution with high risk to the environmental and human health [2], which is extremely disturbing as their use will continue to increase in the coming years. The United Nations reports chemical production for agriculture doubled up to 2.3 billion tons since 2000 and is projected to again double in 2030. The use of chemicals to control pest, diseases, and weeds for productive agriculture is not an issue, but, rather, the irresponsible and discriminate use is the problem posing the considerable high-risk to the health of human, animal, and the environment [25], hence the need for integrated pest management (IPM) approaches as an alternative to sole use of chemicals must be scaled. In this chapter, the position of IPM in circular agriculture is discussed with ways that IPM could help reduce total dependency on sole and excessive chemicals use and provide enormous benefit to humans, animals, the environment, and farmers.

8.1 Integrated pest management (IPM)

IPM is the use of more than one approach to controlling pests and diseases through a careful consideration of all available and desirable control techniques with appropriate measures that are economically, socially, and environmentally sound to minimizing the development of pest populations [1].

8.2 IPM strategies and techniques and their importance in circular agriculture

In IPM approaches, prevention is the highest priority over intervention/control and requires a good amount of technical knowledge of the local biodiversity and their interactional relationships in that local ecosystem. IPM combines biological, chemical, mechanical, and cultural methods with appropriate measures based on local factors to grow healthy crops with minimal use of pesticides, which reduces or minimizes the pesticides use associated risk to human health and the environment as a sustainable alternative to food production [1].

1. Biological control: the use of beneficial organisms such as bacteria and fungus to control other organisms [25]. All crops grown in the natural environment benefits from this approach

where beneficial biological controlling agents are enhanced naturally to control pathogens and pests. Also, the biological beneficials can be promoted through:

a. Conservation biological control - agroecosystem and/or its immediate surroundings are modified to increase the impact of local beneficial micro/macro-organisms.

b. Augmentation biological control- specific beneficial micro/macro-organisms are released through a mass culture and periodic release to control other organisms. An example is the use of desmodium (*Desmodium tweedyi)* and brachiaria grass (*Brachiaria paspaloides*)to control fall army worm in maize in Africa (Fig. 24).

c. Classical biological control—micro/macro-organisms are introduced permanently into an environment where it does not naturally occur for the purposes of controlling pests and pathogens.

2. Chemical control: the use of conventional agrochemicals to control pests and disease.

3. Physical or mechanical control: the use of barriers such as fences and net to prevent pests from getting access to host plants.

4. Cultural control: using regular on-farm practices such as regenerative agricultural practices to make the environment unconducive for pests and diseases to thrive, etc. Example desmodium "tick-clover" (*Desmodium tweedyi)* and brachiaria grass (*Brachiaria paspaloides*) were used to make the environment unconducive for fall army worms and stemborers African maize fields (Fig 24) [26].

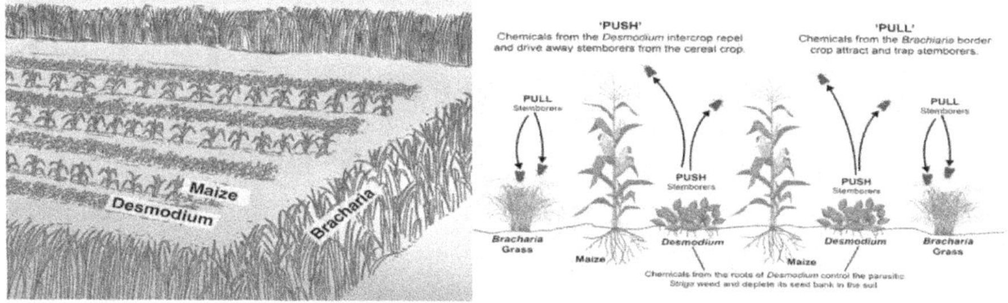

Figure 24: Application of the "PUSH" and "PULL" approach to control fall army worms in Africa as a biological and cultural practice. Source: International Centre of Insect Physiology and Ecology (icipe) and Keele University, United Kingdom. [26].

In a nutshell, IPM is more of a dynamic process that allows growers to use an ecologically friendly approach encouraging a vast range of the best pest control options (biological, cultural, physical, and chemical) available to them with economic, environment, and social considerations that makes sense at the local level. For example, the use of the PUSH and PULL method in Africa to control stemborer and fall army worms (Fig. 24) came with added benefits, such as increased soil fertility, boosted livestock production, increased soil moisture, and more resilience of the local community and especially women against climate change [26]. IPM is based on ecology, the concept of ecosystems and the goal of sustaining ecosystem functions that fall right into the principles and context of circular agriculture. The use of these IPM approaches aims at eliminating one or two of the factors (host, environment, and pathogens/ pests) that makes it possible for pests/ diseases to exist and thrive through the promotion of healthy crops in healthy agro-ecosystems with high biodiversity, hence encouraging natural pest control mechanisms.

8.3 The importance of IPM in circular agriculture

IPM as a circular agricultural practice has great positive impact on food production, human health, farmers' livelihood, and the environment. Below is a list of more importance of IPM;

1. Builds ecosystem services such as natural pest predators, helping control other pests.
2. Protect pollinators and other beneficial organisms for resilient biodiversity.
3. Increases farm productivity and food availability by reducing pre- and post-harvest crop losses and cost of associated synthetic chemicals.
4. Supports safe food for all by reducing pesticide residues in food, feed and fiber, and environment.
5. Enhances ecosystem services and seeks to maintain the natural crop's ecosystem balance that have more benefits beyond the farm boundaries.
6. Conserves the underlying natural resource base such as soil, water, and biodiversity and makes farming resilient against climate change.
7. Enhances ecosystem services such as pollination of crops, increasing carbon for healthy soils, and providing habitat for diversity of species.
8. Increases farmers' livelihood with more income for growers by reducing production costs through reduced levels of pesticide use. For example, in Kenya, the PULL and PUSH system is estimated to add surplus of US$ 72-73 million as economic benefits. [26]
9. Increase quality of food produce due to less residue gives high

and better prices and increases farmers' overall profitability.

10. Strengthens farmer knowledge for a better stewardship of the environment.

11. Increases farmer knowledge of ecosystem functioning adapted to their local context and climate change.

12. Gives growers full control of their farm operations with responsibility and purpose that increase their resilience and sustainability of their business for future generations.

13. Approaches also reduce greenhouse gas emissions through the reduction of agrochemical use and carbon sequestration from the regenerative practices implementation on the farm.

8.4 Step-by-step approach to implementing IPM in circular agriculture systems

1. Understand your environment: Have a good understanding of your farm operation and environment with clear understanding of both potential biodiversity (both beneficials and harmful pests and pathogens).

2. Focus on your soil health: Know your soil health status and the abilities of your soils to support healthy plant growth with a clear long-term plan for soil health improvement. Use regenerative practices such as cover crops, crop rotations, mixed cropping, compost, conservative practices, drainage, mixed farming, etc. to ensure your soils are healthy with the right biodiversity, especially beneficial microbes [1].

3. Plant selection: Based on your soil health and biodiversity status of your farm environment, implement crop rotation with plant selection that will make your field unaccommodating to

harmful pest and diseases while improving the diversity of beneficial organisms. Avoid susceptible species to potential pests and diseases that are a threat to your farm and use resistant plant and animal varieties if available.

4. Plant nutrition: Use the 4R's to ensure plant have the right fertility required throughout the growing season [1]. Also, regularly do plant tissue sampling for nutrition analysis to ensure your plants are well-nourished all time for maximum health.

5. Routing scouting: Do regular scouting to help a identify potential pest and diseases threats at the very early stages for early control in situations they could cause economic damage.

6. Use target chemical control as the last option at the right rate, method, and timing to augment the available biological, cultural, and physical methods.

CHAPTER 9

The Role of Animal Husbandry (Livestock Production) in Circular Agriculture

Animal husbandry is a branch of agriculture that deals with animals' production or raising animals for meat, fiber, milk, or other products. It includes the day-to-day farm activities such as breeding, feeding, disease control either in a complete open field (free range), complete enclosed housed (intensive system), or in between the open and closed system (semi-intensive). The branches of animal husbandry are;

1. Dairy (raising animals such as cattle for their milk as the main product)
2. Meat (raising animals, such as pigs, sheep, goats, and cattle, for their meat)
3. Poultry (raising birds such as chicken, guinea fowls, ducks, turkey, etc. for their eggs or meat)
4. Aquaculture (production of aquatic animal such as fish for either food or other uses)
5. Insect (production of insects such as crickets for food and other uses)

The increasing population correlates with increasing demand for animal food production such as beef, dairy, poultry, and pork around the globe as represented in Table 3. This tells how important livestock production is to food security.

Table 3: Global meat production trends from 2019 to January 2023

1,000 Metric Tons (Weight Equivalent)	2019	2020	2021	2022 Oct	2023 Jan
Total beef and veal production	58,527	57,658	58,366	59,372	59,206
Total pork production	101,030	95,759	107,607	109,846	114,086
Total chicken meat production	97,309	99,257	100,510	100,931	102,94

Adopted from Global Market Analysis by United States Department of Agriculture (USDA) Foreign Agricultural Service reported on 8 January 2023

The Global Market Analysis by the United States Department of Agriculture (USDA) Foreign Agricultural Service clearly shows that there is an increasing production of beef, pork, and poultry meat across the world (Table 3). In the United States alone, sales from animal products such as livestock, dairy, and poultry are more than half the national agricultural cash receipts, which exceeded $160 billion per year since 2015. This makes animal production one of the biggest industries in the world with impact on diverse sectors of our rural and urban economy, health of society and the environment, and livelihood of many small- and large-scale farmers. It is clear at this point that the livestock sector is one of the pillars for global food system with critical contribution to poverty reduction, food security, and agricultural economic development.

According to the Food and Agricultural Organization (FAO) of the United Nations (UN), livestock contributes to 40 percent of agricultural output in developed countries and 20 percent in developing countries with livelihood support to about 1.3 billion people worldwide, biodiversity increase, and carbon sequestration [1]. Projections indicate the livestock sector is expected to see about 50 percent growth by 2050 and will be a major source for worldwide protein (34 percent) supporting while serving as a livelihood source to about 600 million poor, small farmers in developing countries [27].

The positive contributions of livestock to our societies are not without any challenges. For example, the livestock sector is currently projected to emit 7.1 GT of CO2-equivalent per year, which is equivalent to 15 percent of human-induced greenhouse gas emissions [28]. This challenge gives us the opportunity to increase efficiencies in livestock supply chains is key to GHG emissions reduction for the sector moving forward, especially in the context of circular economy. This chapter discusses the role of livestock production in circular agriculture as a sustainable solution in the major role in sustainable food systems. The discussion will focus on manure as a critical source of natural fertilizer, weed and pest control, carbon sequestration and emission reduction, biodiversity the sector adds to our food systems, and the use of livestock as draft animals can help boost productivity in regions limited ability to use machines.

9.1 Importance of livestock in circular agriculture (CA)

1. Biodiversity increases in our food systems: Increasing biodiversity is one of the principles of CA. Several animal species are raised on the farm at small- and large-scale across the globe, including cattle, goats, pigs, poultry, sheep, insects, fish, etc. (Fig. 25 a and b). The sector adds to the species diversity of the world with many benefits in relation food resilience, environmental health, incomes to growers, and source of recreation [1].

Figure 25: *Geese used for weeds and pest control in vineyard (a-Clear Creek Vineyard, Arizona, U.S.A.) and cattle feeding on pastureland (b-University of Arizona Cracchiolo station, Arizona, U.S.A.) (pictures by Dr. Isaac K. Mpanga)*

2. *Nutrient cycling and recycling of waste:* Closing nutrient gaps involve efficient nutrient cycling and recycling using the waste and by-products on the farm and industry (Fig. 3). In the conventional agricultural world, these waste ends on a landfill with more environmental issues such as emissions and water pollution. However, in a circular agriculture system, the waste is a raw material for feed animals either in its raw form or processed as feed and ration for the animal (Fig. 3). Also, the waste from the livestock production and fields becomes fertilizer resources to nourish soil health and to make soils more productive at a lesser cost.

3. *Weed and pest management:* Livestock can be used strategically to manage pest and weeds on the farm. Examples are mixed farming (raising arable crops and animals on the same land either in rotation or concurrently) and agroforestry (integration of agricultural tree crops with animal production with mutual

benefits to both the animals and the tree crops). While the livestock feed extra biomass from the plants, on the weeds, and pests, they return manure as additional benefit to the crops (Fig. 25a and b).

4. *Improve farmers' and community's livelihood:* Livestock are important to our communities, especially underserved communities in developing countries. According to FAO of UN, livestock is estimated to be a source of livelihood to about 600 million small-scale farmers serving them with food, income, and as a store of wealth, collateral, or a safety net in times of need [29].

5. *Carbon sequestration:* Livestock production systems have great potential in carbon capture and storage through good pasture management that is mostly without tillage. In a 20-year multi-species pasture rotation improved soil carbon and other soil health indicators and sequestered an average carbon of 2.29 Mg C ha^{-1} yr^{-1}, which reduced net greenhouse gas emissions by 80 percent [29]. Proper care of pasturelands for livestock production will lead to drastic reduction in associated greenhouse gas emissions through the compensation in carbon sequestration, so our focus should be on best practices.

6. *Food source to the growing population:* The growing population need to be fed, and livestock is projected to be a major food source (Table 3) and will be a source of protein to about 37 percent of the global population by 2050 [27].

9.2 Best practices to minimize greenhouse gas emissions related to livestock production.

It's a fact that livestock production is one of the main sources of greenhouse gas emissions in agriculture, but with best practices in this sector, the gains would outweigh the negatives.

1. Select CA implementation level that is more productive and environmentally friendly with more local partnerships (Fig. 6 and 9).

2. Pasture management must focus on soil health with multi-species forage that will drive biodiversity for more environmental benefits and carbon capture and storage.

3. Use permanent pasture with no-till practices to reduce tillage-associated emissions, reduce run-off, and improve soil biology.

4. Minimize the transportation and exportation of feed from too far distances by partnering with local producers and manufacturers to use their by-products and biomass.

5. Use the organic fertilizers from the animal to fertilize your pasture fields and other crops used in animal feeds.

6. Reduce methane production in cattle. Microbes are responsible for methane production in cattle, so the ability to reduce these actions is important. Some studies suggest grinding and pelleting forage, adding carbohydrates (less digestible feed) and fat to the diet of cattle will reduce methane production [30].

7. Incorporate manure immediately after application in the fields of use injectors to apply liquid manure.

8. Use innovative and climate-smart integrated livestock systems that are more efficient and biodiverse, such as aquaponics, silvopasture (Fig. 25a) [31], increases food resilience. According to a World Bank report, adoption of silvopastoral systems increased milk production (17 percent), reduced production costs (18.5 percent), and increased average number of cows per

hectare by 23 percent between 2011 and 2018 [28].

9. Form partnerships with local growers and manufacturers at a feasible CA implementation level (Fig. 6) for the exchange of by-products and waste. This will improve the circular system and close the loops and leakages in the system with reduction in waste while improving the efficiency of the systems.

10. Select plant species with less water demands for your pasture establishment, biomass, or grain as feed for your livestock, especially if you are in water-stressed regions.

11. Use wasted food and by-products from the farm, factory, and homes to feed livestock directly or after processing with careful food safety considerations. This eliminates organic materials going to the landfills, but, rather, used them to produce food.

CHAPTER 10

Challenges of Circular Agriculture and the Role of Research, Governance/ Politics, Industry, and Society/ Culture in Scaling

Though CA uses old practices, current innovations, and technology as a climate-smart solution that comes with other benefits, however, more work is required, especially at the local levels to make sure policies, research, industry, farmers, and society are on the same table working in the same direction in synergy that efforts are complemented. This will enhance efficient use of resources and to overcome challenges along the way. In this chapter, challenges facing CA and the role of research, governance, industry, farmers, and society are discussed.

10.1 Challenges to Circular Agriculture Implementations

1. Clarity on what is CA and what is not: The clear definition on CA is still relatively not clearly defined, which makes it difficult to tell what is and what is not.

2. Innovation and knowledge gaps on CA at local levels. Science-based information that will lead policy decisions on CA at local, state, national, and global level are limited, which makes it difficult to implementers.

3. Politics: In some cases, there are no political will for CA partly due to lack of scientific information or no interest in climate issues.

4. Finance: Research and technology requires funding to progress, but as of now, not much funding is dedicated for CA. Partly some industries perceive successful implementation of CA equal taking them out of business.

5. Complexity of farming systems: Farming systems vary widely, and climate and weather difference add another layer of complexity, which increases investments requirements to identify practices and technology most applicable to each locality.

6. Contaminants in waste materials: The use of waste materials for food production always comes with food safety-related issues.

7. Social acceptance (taboos): Culture of a society such as taboos and gender could be barriers to both adoption and scaling CA.

8. Quality control and standards on integration of systems: There are currently limited standards on what CA products should be like in terms of quality.

10.2 Role of Research, Governance/Politics, Industry, and Society/Culture on Circular Agriculture

Many partners (research, government, industry, society, etc.) have profound influence on future success or failure of CA, partly due to passionate interest, misinformation, climate changes, solution hunt, etc. For progress to be made in the right direction, transparency with close collaborations is required, and this must be done with understanding that all of us have roles to play (Fig. 26) and one solution can not fit all contexts. Clear identification of stakeholder and partners at the early stage of each CA initiation and allowing them to play related roles in their abilities and capacities with ownerships are key to making strides in CA implementation as a climate crisis solution. This section elaborates the role of key stakeholders in CA as catalyst for scalability (Fig 26).

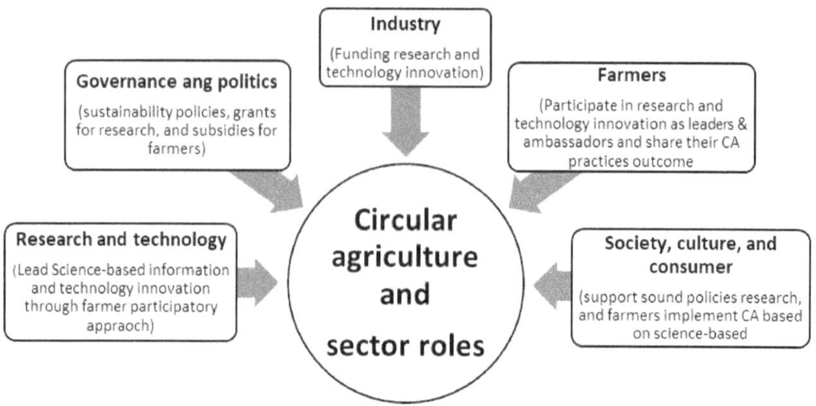

Figure 26: *The role of different sector partners in promoting sound and science-based circular agriculture.*

Research and technology: There are lots of research and technology needs to connect the dots on many fronts of circular agricultural practices to make science-based information more available to farmers, industry, and policymakers for decision-making. According to Helgason, et al (2021) [2] moving towards circular agriculture should not mean returning to past practices, but, rather, a modern way of farming with nature using current scientific advancements, innovations, and new technologies. The use of these science-based approaches, innovations, and technology, should make agriculture more sustainable, productive, eco-friendly, and resilient against climate change. For example, research and breeding can help select crops and animals that are more adaptable to specific CA context, good nitrogen fixing crops, disease and pest resistance, water efficiency, companions in mixed cropping and farming, carbon sequestration, and high-quality crops that would reduce total dependance on external inputs, hence reduce greenhouse gas emissions (Fig. 27).

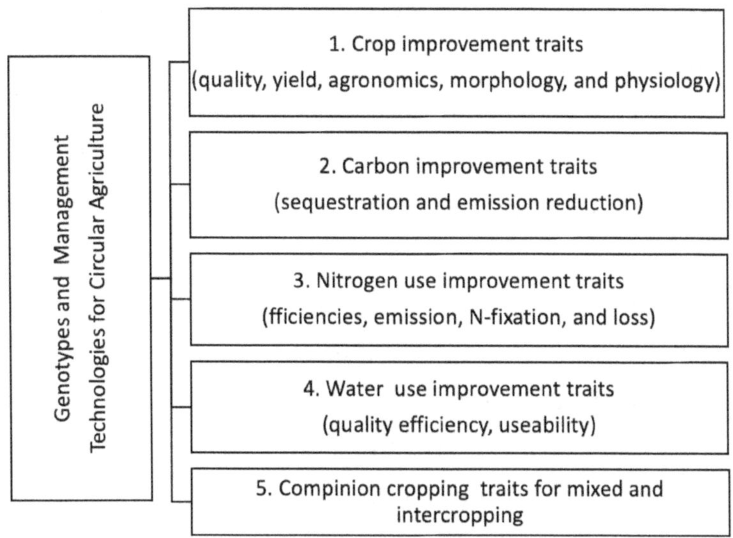

Figure 27: The role of breeding and crop improvement that supports circular agriculture systems through genotypes and management technologies (modified from Messina et al. 2022) [32].

Innovations and technologies such as water treatment would help recycle water, soil mapping, and grid sampling suing GIS would help reduce fertilizer inputs through precision applications, and drones and electric technology could help reduce fuel use on the farm. The use of farmer-participatory research that involves the farmers as stakeholders in the research and technology process are key factors to accelerate these innovations. Farmers would accept research and technological innovations and serve as ambassadors of change if they can own them through direct involvement, continuous engagements, and making them research and innovation leaders in their communities.

Government and politics: Policies and governance could affect scaling of innovative CA positively or negatively, depending on the interest of the government and political interest in CA. For example, a government's subsidy schemes and orientation towards export would favor linear approaches (monoculture) over more circular approaches, such as integrated pest and disease management that leverages biodiversity and crop rotation as a beneficial practice. However, a government with CA interest creates an enabling environment to enable a transition towards circular agriculture through subsidy schemes and trade regulations making use of CA opportunities at internal and external levels [33]. Briefly, public policies placing high value on achieving sustainable use of natural resources would encourage adoption of modern technologies such as drip irrigation, precision agriculture, rainwater harvesting, and crop productivity [2]. Agricultural research can also focus on enabling smallholder farmers to adopt such technologies as well as closing the yield gap between organic and conventional farming.

Industry: Industry has a strong role to play in financing research, innovation, and technology to support CA by either providing finance or inventing innovations and technology needed. For example, tech companies can support technology development such as drip irrigation, multi-species planters and harvesters, composter and compost turners, precision inputs application machines, etc. Also, supply chain companies like food and beverages companies sourcing their raw materials directly from the farmers, could partner with the farmers to recycle the waste in their manufacturing plants as compost or feed for animals. They could support them to implement sustainable regenerative and CA by providing financial incentives and investments and using CA as a benchmark for sourcing. Also, they can work with researchers and universities to provide scholarship schemes to graduate and postgraduate students to work on CA.

Farmer: Farmers are major end users and implementers of CA policies, research, innovation, and technology, hence, getting them involved in the preliminary stages is key to making progress in CA. In this context, the farmer's role includes collaborating with research and technology innovation, acting as owner, leader, and ambassadors to promote implementations. Also, early implementing farmers must be willing to share CA practices' outcome on their farm openly with researchers, other farmers, industry, and government. Also, looking at the current and future competition for farmlands with other growing economic sectors, small- to medium-scale farmers have a major role as we look to scaling these CA systems because most of them are already implementing these CA practices when compared to extremely large farms that are intensive chemical users and that will be the future of farms. The impact these small farms have in our societies and their key positioning in future food production (Fig.28) [7], we must not ignore the small to medium scale growers already implementing these practices but get them involved and collaborate with them for shared learning in the form of farmer-to-farmer knowledge transfer.

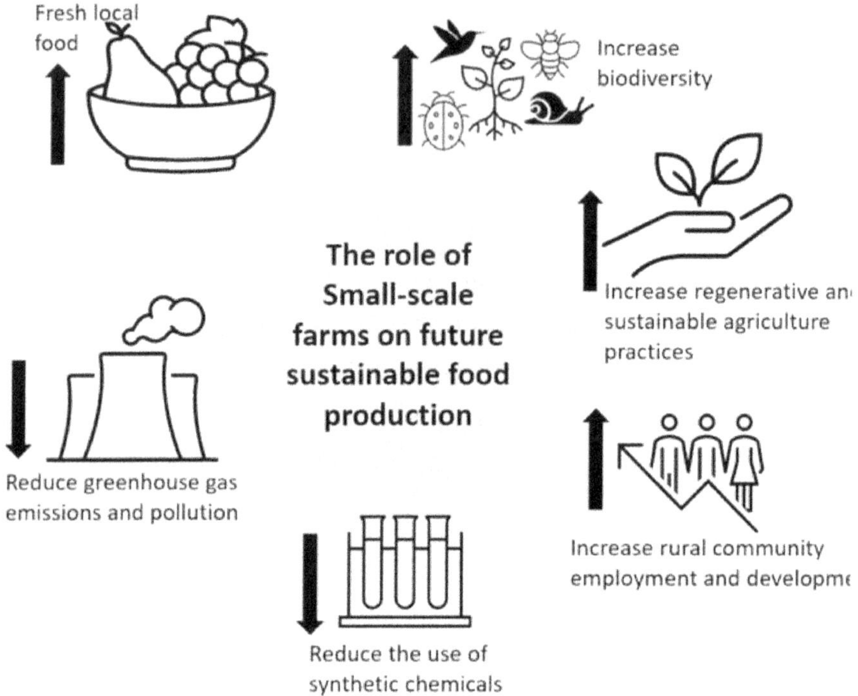

Figure 28. *Small- to medium-scale farms as potentials drivers to circular agricultural economy and future food security [7]*

Society, culture, and consumers: Society and consumers are end-users of CA products such as fresh agricultural and processed products from the farm and factory, respectively. The consumption of these products is influenced by income, culture, personal values, and interest in sustainability. Science-based information available and right delivery approach could shape societal role towards CA positively and otherwise is also true that if we are not able to provide timely science-based information misinformation could also shape societal role to CA negatively. To accelerate CA in society and among consumers, researcher, policies, and technology must always factor in the cultural components of each society to help identify CA practices and levels that are most applicable to each society. Also, use influencers and leaders of each society as ambassadors of your program. In this context, the farmers could still support in those leadership roles if they have good relations, reputation, and live in those communities.

CHAPTER 11

Benchmarks, Evaluation, and Upscaling Circular Agricultural Systems

Benchmarks and evaluation of Circular Agricultural Systems: The ability to set benchmarks and baselines with understanding of where we are today and where we are going are the most basic elements that help and guide our evaluation process to determine progress. It is particularly important to fit circular agriculture evaluation process into this scenario with understanding of the farm, environment, and quality of life today on a linear agricultural system as a baseline, and setting goals on where the farm will be in the coming years after CA implementation.

11.1 How to set benchmark and baseline for a CA system

Set benchmarks around your CA goals, which may differ slightly based on each farm, location, and the level of CA applicable. It is important that all benchmarks have basic CA principles discussed in Chapter 2. The most useful benchmarks will be already existing CA farms that can be a model, however, not many well-established CA farms exist for this purpose. After setting the benchmark, the next step is to measure baseline data based on the current system and/or first year of CA that will help track progress in the coming years through periodic data collection of the same parameters. The baseline data measured must be the same as the periodic data collected after CA transition to facilitate good evaluation of system.

Example: Gunter in Germany has a linear maize monoculture farm operation that he wants to transition into CA with the goal of improving on the eight matrices in Table 4 within 10 years. He is partnering with a cattle farmer and a maize manufacturing plant at the community level of CA. The baseline data to collect in this case would be as listed in Table 4b.

Table 4: Example of CA goals, the baseline data required, and the tools for data collection for evaluation and assessment of progress

a. Suggested CA goals	b. Baseline data	c. Suggested tools for data collection
1. Carbon sequestration	a. amount of carbon in the soil	I. lab test (Haney test), online calculator such as Cool Farm Tool
2. Greenhouse gas (GHG) reduction	b. the amount of GHG produces on the farm	II. Cool Farm Tool, COMET-Farm Tool
3. Nitrogen and phosphorus recycling	c. Nitrogen and Phosphorus levels in the soil	III. Lab test of soil for residual nitrogen and phosphorus
4. Water use efficiency	d. amount of water used on the farm	IV. models, manual calculation

5.	Biodiversity	e.	species diversity	V.	Manual counting of species, Cool Farm Tool
6.	Environmental quality improvement (water quality)	f.	leaching, erosion, and water quality	VI.	lab testing for water bodies mineral (nitrate, phosphorus, microbes, and heavy metals) concentrations
7.	Zero farm waste	g.	amount of waste burnt or disposed	VII.	Taking manual records on waste amount
8.	Wellbeing of farmers, workers, and consumers	h.	the wellbeing of family and workers	VIII.	Taking manual records on family and workers health and quality of life

11.2 How to measure and evaluate progress on CA

Though measuring and evaluating CA may vary among implementers based on goals and objectives and the hierarchical level at which they implement CA. However, it must be based on the general CA principles and individual goals for implementing CA. For example, once the goals and benchmarks are set and baseline data collected accurately, use science-based guidance to determine tools, methods, timing, frequency of data collection, and analysis (Table 2). If possible, use the same approach for the baseline data collection for all subsequent data collection to keep consistency at a level that would make year to year (or any regular period) comparisons possible and easy. For lab analysis that involves soil samples, recommendations are to sample at the same time of the year, same soil depth, and use the same lab for all the analysis.

11.3 Upscaling CA innovations

Based on Mpanga's model for circular agriculture implementation (Fig. 6), starting from the level that is closest to the farmer's practice and advancing up the pyramid by the years and experiences and gaining rapport among the communities and building network will guarantee success at more efficient and productive levels of CA. Secondly, recognizing the role of all stakeholders and gaining their trust to work together as a team from the very beginning with respect is key to successfully scaling CA, and it must start from the local communities (grassroots). According to Biancha, et. Al. (2020), scaling innovation such as CA is a social process with complexity and factors such as division of power, social stigmas, and quality standards checks could hinder speed of progress. Model farmers ('early adopters') are potential drivers for innovation progress and are crucial key partners in upscaling, due to the respect bestowed on them by their communities [33]. Also, both women and men, especially in leadership role at the family and community levels, are influencers for upscaling innovation. Women are more influential at the family level as leaders in family decision-making but cultural barriers and limited external connections limits their ability to influence decisions externally. Men, on the other hand, could influence more external decisions, due to their scope of influence, especially in the context of developing countries. To propel CA innovations' acceptance at an optimal level, there must be awareness and understanding on the role of culture and social norms, gender, community leadership, model farmers (early adopters), language, government policies, industry, regular training and education, and quality control standards [33].

Bibliography

[1]" FAO. The State of the World's Biodiversity for Food and Agriculture, J. Bélanger & D. Pilling (eds.). FAO Commission on Genetic Resources for Food and Agriculture Assessments. Rome. 572 pp. 2019. (http://www.fao.org/3/CA3129EN/CA3129EN.pdf)

[2]"UN/DESA Policy Brief #105: Circular agriculture for sustainable rural development | Department of Economic and Social Affairs," *www.un.org*, 2019. https://www.un.org/development/desa/dpad/publication/un-desa-policy-brief-105-circular-agriculture-for-sustainable-rural-development/

[3]R. Naidu *et al.*, "Chemical pollution: A growing peril and potential catastrophic risk to humanity," *Environment International*, vol. 156, no. 156, p. 106616, Nov. 2021, doi: https://doi.org/10.1016/j.envint.2021.106616.

[4]M. Anas *et al.*, "Fate of nitrogen in agriculture and environment: agronomic, eco-physiological and molecular approaches to improve nitrogen use efficiency," *Biological Research*, vol. 53, no. 1, Oct. 2020, doi: https://doi.org/10.1186/s40659-020-00312-4.

[5]EPA. "Estimates of generation and management of wasted food in the United States in 2018," 2020. Available: https://www.epa.gov/sites/default/files/2020-11/documents/2018_wasted_food_report-11-9-20_final_.pdf

[6]"Circular agriculture: a new perspective for Dutch agriculture," *WUR*, Sep. 13, 2018. https://www.wur.nl/en/show/circular-agriculture-a-new-perspective-for-dutch-agriculture-1.htm

[7]I. K. Mpanga, U. K. Schuch, and J. Schalau, "Adaptation of resilient regenerative agricultural practices by small-scale growers towards sustainable food production in north-central Arizona," *Current Research in Environmental Sustainability*, vol. 3, no. 3, p. 100067, 2021, doi: https://doi.org/10.1016/j.crsust.2021.100067.

[8]A. Muscio and R. Sisto, "Are Agri-Food Systems Really Switching to a Circular Economy Model? Implications for European Research and Innovation Policy," *Sustainability*, vol. 12, no. 14, p. 5554, Jul. 2020, doi: https://doi.org/10.3390/su12145554.

[9]A. M. lecture to: ir Jbm. C. form, "Circularity in agricultural production," *WUR*, Jan. 25, 2019. https://www.wur.nl/en/newsarticle/circularity-in-agricultural-production-1.htm (accessed Feb. 10, 2023).

[10]A. Jurgilevich *et al.*, "Transition towards Circular Economy in the Food System," *Sustainability*, vol. 8, no. 1, p. 69, Jan. 2016, doi: https://doi.org/10.3390/su8010069.

[11]J. Poore and T. Nemecek, "Reducing food's environmental impacts through producers and consumers," *Science*, vol. 360, no. 6392, pp. 987–992, Jun. 2018, doi: https://doi.org/10.1126/science.aaq0216.

[12]D. Marinova and D. Bogueva, "Circular Agriculture," *Food in a Planetary Emergency*, vol. 2, no. 6, pp. 75–92, 2022, doi: https://doi.org/10.1007/978-981-16-7707-6_5.

[13]UNEP, "UNEP Food Waste Index Report 2021," *UNEP - UN Environment Programme*, Mar. 04, 2021. https://www.unep.org/resources/report/unep-food-waste-index-report-2021

[14]S. K. Joshi and A. K. Gauraha, "Global biofertilizer market: Emerging trends and opportunities," *Trends of Applied Microbiology for Sustainable Economy*, pp. 689–697, 2022, doi: https://doi.org/10.1016/b978-0-323-91595-3.00024-0.

[15] I.K. Mpanga, H. Sserunkuma, R. Tronstad, M. Pierce, and J.K. Brown (2022). Soil health assessment of three semi-arid soil textures in an Arizona vineyard irrigated with reclaimed municipal water, Water 14, no. 18: 2922.

[16]I. K. Mpanga, "Fertilization Strategies to Improve the Plant GrowthPromoting Potential of Microbial Bio-Effectors.," Doctoral Dissertation, UNIVERSITY OF HOHENHEIM, 2019. Available: http://opus.uni-hohenheim.de/volltexte/2020/1750/pdf/Doctoral_Thesis_Isaac_Mpanga_approved.docx.pdf

[17]I. Mpanga *et al.*, "The Form of N Supply Determines Plant Growth Promotion by P-Solubilizing Microorganisms in Maize," *Microorganisms*, vol. 7, no. 2, p. 38, Jan. 2019, doi: https://doi.org/10.3390/microorganisms7020038.

[18]I. Mpanga, H. Dapaah, J. Geistlinger, U. Ludewig, and G. Neumann, "Soil Type-Dependent Interactions of P-Solubilizing Microorganisms with Organic and Inorganic Fertilizers Mediate Plant Growth Promotion in Tomato," *Agronomy*, vol. 8, no. 10, p. 213, Oct. 2018, doi: https://doi.org/10.3390/agronomy8100213.

[19]"CORDIS | European Commission," *Europa.eu*, 2023. https://cordis.europa.eu/project/id/312117/reporting (accessed Feb. 11, 2023).

[20]A. S. Pascaris, C. Schelly, L. Burnham, and J. M. Pearce, "Integrating solar energy with agriculture: Industry perspectives on the market, community, and socio-political dimensions of agrivoltaics," *Energy Research & Social Science*, vol. 75, no. 75, p. 102023, May 2021, doi: https://doi.org/10.1016/j.erss.2021.102023.

[21]C. Dupraz, H. Marrou, G. Talbot, L. Dufour, A. Nogier, and Y. Ferard, "Combining solar photovoltaic panels and food crops for optimising land use: Towards new agrivoltaic schemes," *Renewable Energy*, vol. 36, no. 10, pp. 2725–2732, Oct. 2011, doi: https://doi.org/10.1016/j.renene.2011.03.005.

[22]H. Dinesh and J. M. Pearce, "The potential of agrivoltaic systems," *Renewable and Sustainable Energy Reviews*, vol. 54, no. 54, pp. 299–308, Feb. 2016, doi: https://doi.org/10.1016/j.rser.2015.10.024.

[23]P. Santra, P. C. Pande, D. Mishra, and R. Singh, "Agri-voltaics or Solar farming: the concept of integrating solar PV based electricity generation and crop production in a single land use system," *Ijrer.org*, 2023. https://www.ijrer.org/ijrer/index.php/ijrer/article/view/5582/0 (accessed Feb. 11, 2023).

[24]D. Mavani, P. Chauhan, and V. Joshi, "Beauty of Agrivoltaic System regarding double utilization of same piece of land for Generation of Electricity & Food Production," *International Journal of Scientific & Engineering Research*, vol. 10, no. 6, 2019, Accessed: Feb. 11, 2023. [Online]. Available: https://www.ijser.org/researchpaper/Beauty-of-

Agrivoltaic-System-regarding-double-utilization-of-same-piece-of-land-for-Generation-of-Electricity-Food-Production.pdf

[25] E. T. G. Niekawa *et al.*, "The microbial role in the control of phytopathogens—an alternative to agrochemicals," *Microbiome Stimulants for Crops*, pp. 159–177, 2021, doi: https://doi.org/10.1016/b978-0-12-822122-8.00015-7.

[26] I. S. Sobhy *et al.*, "Bioactive Volatiles From Push-Pull Companion Crops Repel Fall Armyworm and Attract Its Parasitoids," *Frontiers in Ecology and Evolution*, vol. 10, no. 10, Apr. 2022, doi: https://doi.org/10.3389/fevo.2022.883020.

[27] "Publication card | FAO | Food and Agriculture Organization of the United Nations," *www.fao.org*, 2019. https://www.fao.org/publications/card/en/c/CA7660EN/ (accessed Feb. 11, 2023).

[28] The World Bank, "Moving towards sustainability: The Livestock Sector and the World Bank," *World Bank*, 2019. https://www.worldbank.org/en/topic/agriculture/brief/moving-towards-sustainability-the-livestock-sector-and-the-world-bank

[39] J. E. Rowntree *et al.*, "Ecosystem Impacts and Productive Capacity of a Multi-Species Pastured Livestock System," *Frontiers in Sustainable Food Systems*, vol. 4, no. 4, Dec. 2020, doi: https://doi.org/10.3389/fsufs.2020.544984.

[30] M. Jones, "Ways to Reduce Methane Production in Cattle," *UNL Beef*, Feb. 2014. https://beef.unl.edu/reduce-methane-production-cattle

[31]I. Mpanga, J. Allen, and U. Schuch, "Agroforestry as a sustainable ancient agriculture practice: potential for small-scale farmers and ranchers in dry regions," 2021. Available: https://extension.arizona.edu/sites/extension.arizona.edu/files/pubs/az1918-2021.pdf

[32]C. D. Messina *et al.*, "Crop Improvement for Circular Bioeconomy Systems," *Journal of the ASABE*, vol. 65, no. 3, pp. 491–504, 2022, doi: https://doi.org/10.13031/ja.14912.

[33]F. Bianchi, C. Van Beek, D. De Winter, and E. Lammers, "Opportunities and barriers of circular agriculture insights from a synthesis study of the Food & Business Research Programme," Mar. 2020. Available: https://www.nwo.nl/sites/nwo/files/documents/1.%20 Circular%20agriculture_full%20paper.pdf

About the Author

Isaac Kwadwo Mpanga (PhD) was born in Ghana (West Africa), married to his beautiful wife (Sabrina Campbell Mpanga), and they have a smart son (Josiah Taasun Mpanga), who keeps them active. He loves going to church, hiking, taking road trips, going to parks, taking walks in their neighborhood with family, biking, and playing toddlers' football and basketball with his son.

Dr. Mpanga grew up on the farm as a farmer, trained and practices as a teacher, researcher, and consultant with lifetime experience in the sustainable agricultural space. His research interest focuses on soil health, sustainable farming and cropping systems, and regenerative agriculture, and he has over thirty peer-reviewed scientific journal articles, a book chapter, factsheets, posters, and conference presentations published in high-impact and reputable journals as lead author. He holds a PhD in agricultural sciences and MSc in crop sciences, both from the University of Hohenheim in Germany. He also has a BSc degree in agriculture education from the University of Education, Winneba (Ashanti Mampong) and associate degree in education from the University of Cape Coast in Ghana. His high school education was at Krachi Secondary Technical School at Kete-Krachi, and his basic education was in a small town call Borae # 2, where he grew up.

Dr. Mpanga currently works at PepsiCo North America business unit as R&D Associate Manager (Agro Sustainability and Best Practices), where he leads and supports regenerative agriculture and sustainably sourcing program execution for Frito Lay's potato growers under the PepsiCo positive goal. His work at PepsiCo focuses on reducing greenhouse gases emissions in potato fields, increasing biodiversity, enhancing soil health, efficient water use, and improving farmers' livelihoods.

Before he joined PepsiCo, he was a faculty (Area Associate Extension Agent) at the University of Arizona, where he created the Commercial Horticulture and Small Acreage program for Yavapai and Coconino Counties to improve local food production and food security resilience through implementation of sustainable agricultural practices through research and extension programming that delivers science-based information to stakeholders. In this role, he was able to lead research proposals as the project director with over USD $2 million funding from United State Department of Agriculture (USDA) under the National Institute of Food and Agriculture (NIFA) organic transition grant and Natural Resources Conservation Services (NRCS) CIG on-farm soil health demonstration grants to support climate-smart research and extension programming for small-farmers in the state and beyond.

Other places he worked are in Germany under the BIOFECTOR Partner-Projects as a research fellow while he was pursuing his PhD on the same project where he focused on fertilization strategies to improve the plant growth promotion of biofertilizers to help European agriculture cut down in synthetic fertilizer use. In Ghana, he also taught as an intern at the University of Energy and Natural Resources in Sunyani and as a teacher at Ghana Education Services (GES) for many years.

Dr. Mpanga's inspiration to writing this book came from the experiences gained on the farm and professional training. He wanted to share expertise on circular agriculture that do not fit into the traditional scientific publication space where most of his work is documented as scientific publications that may either not be accessible to all who need them, or they are too complex to use by other people who are not scientists. The book gives a comprehensive guide to circular agriculture as a climate solution for all interested in climate solutions and sees agriculture as a major driver. The book discusses principles and practices of circular agriculture and the role of a hierarchical model implementation he developed and stakeholders such as farmers, researchers, manufacturers, government, and community on scalability. Get your copies with your teammates and friends and dive into the circular agriculture guide on how agriculture could drive climate solutions.